PERRY'S LAKE ERIE FLEET

PERRY'S LAKE ERIE FLEET

After the Glory

DAVID FREW

THE
History
PRESS

Published by The History Press
Charleston, SC 29403
www.historypress.net

Front cover: *Prelude to Battle*, Peter Rindlisbacher.
Back cover: Oliver Hazard Perry, Jerry Skrypzak Collection.

First published 2012
Manufactured in the United States

ISBN 978.1.60949.610.4

Library of Congress Cataloging-in-Publication Data
Frew, David R.
Perry's Lake Erie fleet : after the glory / David Frew.
p. cm.
Includes bibliographical references and index.
ISBN 978-1-60949-610-4
1. United States--History--War of 1812--Naval operations. 2. Great Lakes Region (North America)--History, Naval--19th century. 3. Lake Erie, Battle of, 1813. 4. Perry, Oliver Hazard, 1785-1819. 5. United States. Navy--History--War of 1812. I. Title.
E360.F74 2012
973.5'254--dc23
2012007764

CONTENTS

CONTENTS

Acknowledgements

The three primary persons whose artistic contributions were essential to this project were Isaac Smith, chart and map illustrator from the Mercyhurst University Graphic Art Department; my friend and coauthor Jerry Skrypzak, photographer; and Courtney Sullivan, editor and technical assistant.

Several others made important contributions, including John Baker, photographer; Harry B. Barrett; Ian Bell and the Port Dover Harbour Museum; Linda Bolla; David Brunelle of the Southern Georgian Bay War of 1812 Commission; Hannah Cassilly; Alyson Cummings and the Erie County Historical Society; Laural Finney; Dr. John C. Fredriksen, who compiled the diary of Usher Parsons; Mary Ann Frew; Dr. William Garvey and the Jefferson Educational Society of Erie; Dr. Gil Jacobs of Mercyhurst's Organizational Leadership Graduate Program; Tom Leonardi; Mike Lynch and the Erie Yacht Club; Mercyhurst University and its library resources; the Niagara League; Hilary Parrish; Port Dover Board of Trade; Captain Walter Rybka from the brig *Niagara*; Dave Stone; Kathleen Trainor and Huronia Historical Parks; Dr. Robert Wall; Terry Walsh; the Wasaga Beach Provincial Park; and my hostage 2011–12 graduate students who listened to interminable War of 1812 stories, read trial chapters and made thoughtful recommendations. Special thanks is also owed to my grandchildren, who sometimes endure a distracted Grampa who is writing these stories: Abigail, Colin, Eden, El, Hudson, Jordi, Noah and Phinaus.

Introduction

GENESIS OF THE
BOOK PROJECT

History is an enlargement of the experience of being alive.
—historian David McCullough, discussing his book Brooklyn Bridge

I have lived most of my life in Erie, Pennsylvania, a delightful town on
the shores of Lake Erie, and for much of that time I have found myself
serving as an apologist. Most of my out-of-town colleagues who have not
experienced Erie recall ugly television images of a terribly polluted lake,
punctuated by the stench of dead fish. Or they remember the infamous
burning of Cleveland, Ohio's Cuyahoga River in 1969. That event
may have been the signature moment that shifted resources toward the
environmental renewal of Lake Erie and those efforts have paid great
dividends. Friends from other cities also ask about the demise of the city's
manufacturing base and the high-paying union jobs that once drove the
regional economy. They wonder if Erie has become a pocket of abject
poverty within the decaying Great Lakes rust belt, battered by frightful
weather, winter snow and lake effect storms.

The actual city that I have experienced could not be more different
from those negative stereotypes. Erie lies on the south shore of a vastly
improved Great Lake, where it enjoys an amazingly moderate climate.
Its lakeshore location creates a micro-climate, with extended fall seasons
that have allowed for the development of a delightful grape-growing and
wine industry. Lots of snow falls each winter, but most of it melts quickly
because of our temperate location. Fortunately, just enough sticks to the

hills southeast of the city to powder the ski slopes. And of the missing manufacturing economy, I always tell my out-of-town friends that Erie is a maritime town that accidentally became a manufacturing center for a brief moment in anthropological time.

The signal event in Erie's history took place when the United States decided to open a naval base here in 1813. The sheltered harbor behind the peninsula, which protects Erie from the open lake, provided protection from the British when the United States Navy built its Great Lakes naval fleet here in 1813. That fleet subsequently met and defeated the British at the Battle of Lake Erie, helping to end the War of 1812. During this remarkably short wartime period when Erie was perched on the leading edge of America's western frontier, its population accelerated from a few hundred to a few thousand people almost overnight. It was, clearly, the United States Naval Base at Presque Isle that was responsible for this growth.

Having a naval base in the tiny frontier town of Erie was much like the addition of a NASA station would be in a modern city. In those days, sailing ships were much like today's rocket ships that depart earth and head toward unimaginable distant places. Nineteenth-century sailing ships represented a mystical connection to the big world, and Erie quickly became the place where they were built and berthed. Shipbuilders, sailors, soldiers, merchants and others flooded into the town, which was located at the best strategic location on the most important of the Great Lakes. Amazingly, from the context of modern society, tall ships represented the fastest form of transportation on the planet during those early days, and the romance of watching sails appear or disappear across a water horizon fueled the excitement and imagination of the town's populace.

While the War of 1812 was relatively short and Erie's naval base was only an active military center for a few years, the influence of that time during the town's formative years was significant. There had been merchant shipping on Lake Erie for a decade preceding the war, but the influx of large military warships helped demonstrate the potential for commercial shipping on the upper (above Niagara Falls) Great Lakes. Within a decade after the war, Lake Erie was bustling with merchant schooners, and the first steamships had appeared. Leftover sailors and shipbuilders from Erie's naval days transferred their skills to the evolving maritime economy and helped Erie (and other Great Lakes cities) shift toward a new commercial maritime identity. By the pre–Civil War era, Erie was the home port of the largest fleet of steamships in the United States, and its sheltered harbor had become one of America's busiest inland ports.

Contemporary aerial view of Erie's Presque Isle Bay, former home of Perry's War of 1812 fleet. *Photograph by Jerry Skrypzak.*

North America's canal era further helped to fuel Erie's maritime identity. The 1829 creation of Ontario's Welland Canal, which bypassed Niagara Falls, and Dewitt Clinton's 1825 Erie Canal from New York City to Buffalo brought even more traffic to Lake Erie. By the middle of the 1800s, Lake Erie had become the busiest waterway in the world, and Erie was strategically located in the center of this important Buffalo to Detroit trade route. To participate in the new canal era, local businessmen built a local canal to the south from Erie to Pittsburgh. The Erie Extension Canal attracted a wave of passenger and freight traffic from Lake Erie into the city.

As Great Lakes ships shifted from sail to steam, Erie forged another related maritime identity. By the mid-1800s, it had become the steam and boiler capital of the world, providing engines and running gear for the vessels that were plying the inland seas. Then the city began to develop an infrastructure of machine shops and foundries that were focused on ships, shipbuilding and maritime maintenance. The emerging expertise in steam engines and boilers also advanced the city's commercial fishing industry. If Erie did have an identity as a manufacturing town, it was forged by its evolution from naval base to shipbuilding, shipping, transfer station, canal and commercial fishing center.

Like most northeasterners with a love of history, I have made many visits to Gettysburg and other iconic Civil War locations south of here. When anyone visits such a hallowed site, they are immediately transported to the terribly violent War Between the States and propelled into a historical time capsule. My time at Gettysburg and its battlefields, while inspirational and filled with emotion, has also helped me to understand something about Erie. For the citizens of this city (and other Great Lakes cities), the War of 1812 represents much of the same connection and emotion that the Civil War does for citizens and visitors at Gettysburg. The seemingly obscure War of 1812, Erie's war, was a transformational event that forever sealed the identity of this city as a maritime town.

Although the actual Battle of Lake Erie was fought in Ohio, more than one hundred miles west of Erie, that historic event, which marked the first time that our fledgling country challenged Great Britain to a naval contest and captured an entire fleet, is a local historical benchmark. The victorious American fleet returned to Erie in September 1813, bringing its captured British ships and prisoners with it. Then the fleet and its naval personnel remained in residence here. For that reason, locals have always imagined the September 10, 1813 date of the battle—as opposed to the 1812 date of the war itself—as an important marker. In 1913, the city celebrated the 100th anniversary of the battle by raising Oliver Hazard Perry's flagship, the brig *Niagara*, from the bottom of Misery Bay, where it had been scuttled in 1820. The *Niagara* was restored and taken on a triumphant Great Lakes tour that year, and from that eventful day forward, the brig has continued to be a treasured member of the community.

With the 200th anniversary of the Battle of Lake Erie approaching, Erie's Jefferson Society, a local think tank, convened a panel to discuss preparations for the anticipated year 2013 events. Preliminary discussions centered on publications and especially books that might focus on Erie's place in War of 1812 history. The panel discussed the question of "what might be needed to add to the volumes of material that had already been published about the war in general and the Battle of Lake Erie in particular." In searching the available literature, dozens of well-written, detailed volumes covering the battle, its naval tactics and the many questions of how the American fleet actually won the skirmish were uncovered. After reviewing these works, it was generally concluded that more than enough books and articles had already been written about the Battle of Lake Erie itself, the geo-political context of the war, the importance of the fighting that took place near the Great Lakes and the extreme logistical difficulties that hampered the delivery of supplies to out-of-the-way frontier outposts such as Erie.

What had not been written and compiled in a single place, however, was a narrative of the post–Battle of Lake Erie events that helped to keep Erie's naval base open for several more years. The 1814 Campaign in particular was seemingly lost to history. There were articles and side notes, but no comprehensive volume dealing with that history. Perhaps it was because the Battle of Lake Erie seemed to have been such a definitive punctuation mark in the overall war. For participants in the 1814 events, however, the burnings at Dover Mills, the attack on the British-held fort at Mackinac Island and the burning of the supply ship *Nancy* in the Georgian Bay seemed just as important as the Battle of Lake Erie.

Another apparent shortfall in the existing books was a "readable" volume that detailed all of the post-battle history and connected it to the Battle of Lake Erie, itself. Much of what has appeared in naval history and the other volumes dedicated to the War of 1812 is relatively technical material aimed at military strategists, professional historians or naval enthusiasts. Important minutiae such as cannon gauges, ship sizes, material composition of the ships that were built for the battle and other details, while important to technicians, present difficult reading for general readers of this kind of history. In viewing history from 2012, through the long lens of time, the panel decided that it might be possible to put enough chronological and emotional distance between events of the past and new reference points to add to the general understating of earlier events.

With respect to Erie's role in the War of 1812, while there are few remaining local land shrines to the conflict—forts, battlefields or other structures—there is the seemingly eternal brig *Niagara*. Unlike docent tours of Gettysburg's battlefields, where one is left to imagine the events of those days, Erie's most senior citizen, *Niagara*, sits at her berth near the downtown Erie Maritime Museum, constantly reminding us of the events that took place here. And if the dockside presence of *Niagara* is not enough to bring images of the War of 1812 to life, when the brig is untied and slips into Erie's Presque Isle Bay to go sailing, it is impossible for even the most oblivious to fail to notice the presence of this apparition from another time.

It is interesting to note that the newest version of the reconstructed brig *Niagara* has been in the water and sailing for more than twenty years. Thinking of this in the context of elapsed time, Oliver Perry's 1813 experience with the *Niagara* was compressed into less than six months, and Arthur Sinclair's captaincy the following year lasted only five months.

Meanwhile, the current captain of the restored brig, Walter Rybka, has had twenty years to manage the ship, sail it and understand how it would have performed in almost every imaginable circumstance, from traveling back and forth across Lake Erie to ascending the Detroit and St. Clair Rivers and fighting at the actual battle.

The current *Niagara* is equipped with a full complement of cannons, and when it performs its summer maneuvers, its gunnery crew actually fires them, while the sailing crew struggles to maneuver the ship. Watching the difficulties of sailors sailing while gunners load, reload and fire, begs the question of what might have been happening when all of these confusing activities were accompanied by return fire. One of the most devastatingly dangerous aspects of the Great Lakes naval battles as they were fought on hastily built softwood boats with cannons on top decks was the shower of deadly splinters that erupted every time an enemy cannonball impacted a ship. When a cannonball struck a ship, a literal shower of wooden splinters strafed the deck. Each of the individual sharp-edged wooden projectiles that exploded from the impact had two or three times the striking force of an Indian arrow. Enemy cannon strikes were immediately followed by dozens of sailors dropping to the deck in agony, with wooden splinter-arrows piercing their bodies.

Modern sailors, professionals and volunteers, climb *Niagara*'s rig, unfurl her sails and trim the yards to make course, helping observers understand the technical aspects of sailing a tall ship. This experience is so rare in the modern world that contemporary sailors from the United States Navy's *USS Constitution* have come to "learn the ropes." Captain Rybka's experiences aboard *Niagara* help use the long view of time to better understand the early nineteenth-century struggles of Perry and Sinclair, wartime captains of the *Niagara*. And thanks to modern technology, we are able to record and discuss these important technical aspects of sailing and present them to generations of young sailors and historians. Because of the past few decades of sailing aboard *Niagara*, it is likely that we now know more about how the brig might have performed during the War of 1812 than the captains and crew who sailed the original ship into battles. That means that a modern discussion of *Niagara* and its operations can shed light on the history that was recorded almost two hundred years ago by persons with greatly limited shipboard experience.

The expert panel of the Jefferson Society decided that the book that was needed to enhance the 2013 story of Erie's contributions to the coming of age of the United States of America was a readable narrative that would

The reconstructed brig *Niagara* sails in Presque Isle Bay at her home port of Erie. *Photograph by Jerry Skrypzak.*

detail the events that followed the Battle of Lake Erie. And in the telling of those stories, it seemed that the narrative should extend into the present day, while cataloguing the stories of the unique characters who played critical roles in the building and sailing of the fleet at Erie. It should be a story that would celebrate the past by including nuances and insights that have evolved over the years that have passed since the famous battle. The challenge would be to present the characters and the conflicts that dotted the history that followed the battle, and to do so in a way that would be accurate and fair. Given the controversy between Oliver Hazard Perry and Jesse Elliott, who fought over the interpretations of the events at the actual battle, that seemed like a daunting challenge.

One of my continuing inspirations for writing about history is Pulitzer Prize–winning author David McCullough. His book *Brooklyn Bridge* is one of my favorite pieces of history. McCullough has often spoken about using the experience of a lasting physical artifact, in his case the bridge itself, to compel him to write about its history. After walking the iconic New York structure dozens of times, it occurred to him that its lasting physical presence created both a connection to the past and an urgency to tell the stories that informed the bridge project. Perhaps the continuing presence of the brig *Niagara*, which has been an anchor of my life from my earliest recollections, is such an inspiration. My Brooklyn Bridge.

Even with such a physical connection as an inspiration, however, the researcher in me required primary data to serve as a foundation for a narrative. After agreeing to write a story that would bridge the time from the Battle of Lake Erie to the present, I spent months trying to imagine a way to begin. Then, one day, I recalled the diary of Usher Parsons, a young surgeon's mate who was assigned to the fleet at Erie. It was Parsons who had the longest personal experience of any officer aboard the original brig. Fortunately, Parsons left the gift of a diary detailing his time in Erie aboard the ships. Parsons' diary entries were artfully edited into a volume called *Surgeon of the Lakes* in 2000 by historian John Fredriksen, a book that I eagerly read when it was released ten years ago. Thanks to Parsons' observations of the battle itself and his detailed account of the 1814 Campaign that he witnessed firsthand, I had data to work with—a place to begin. Parsons' diary provides one of the most detailed views of the war at Erie, Pennsylvania. His diary entries describe life aboard the ship, the difficult winters at Misery (Little) Bay and the connections between the naval base at Presque Isle and Erie's civilian populace.

To begin the challenge of writing the story of Perry's fleet after the battle, I started with Parsons' diary entries, this time examining them more carefully than I had a decade ago. The narrative that follows within this book begins on the afternoon of September 10, 1813, just after the cannon fire had ended. It transports the reader to the smoldering deck of the *Lawrence*, Oliver Perry's original flagship, where Dr. Usher Parsons is working feverishly to tend to the wounded.

Chapter I

USHER PARSONS' BAPTISM

Cannonballs and Splinters

These [wounded] *were brought down faster than I could attend to them, farther than to stay the bleeding or support shattered limbs with splints and then pass them to the berth deck.*
—*Usher Parsons' diary, September 10, 1813*

Usher Parsons glanced up from his gruesome work. Something had changed. The stench of burnt flesh and blood mixed with gunpowder was the same, and burning bits of rigging and wood were still falling around him. But something was different. His disabled ship, the *Lawrence*, was drifting aimlessly, and over his aching shoulder he could see the outlines of at least four other square-rigged ships. Shrieks of the wounded continued to fill the air, some from Parsons' ship and others from farther away.

Then he knew. The cannon fire had ended. It was 3:15 p.m., and several minutes had passed since a shot had resonated across the surface of the lake. Parsons was working on the damaged deck of the *Lawrence*, alternating between amputating the arms and legs of severely wounded sailors and applying tourniquets, sutures and dressings. Sixty minutes earlier, he had been attending to the crushed arm of Midshipman Taylor Laub in the makeshift one-hundred-square-foot surgical wardroom below decks. Just as he had splinted the wound and recommended that his patient go forward and lie down, a cannonball exploded through an exterior wall, spraying everyone with deadly splinters. Parsons suffered glancing blows in the arm and side of the face from shards of wood. The two sailors who had been

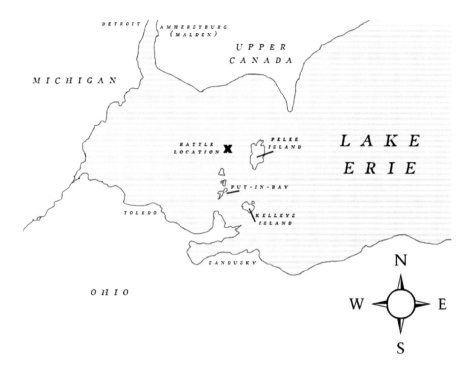

Western Lake Erie and the islands, including Amherstburg and the location of the Battle of Lake Erie. *Illustration by Isaac Smith.*

assisting him fell to the floor, screaming in agony. Splinters had lodged like arrows in their necks and faces.

The larger calamity, however, was Midshipman Laub, whom Parsons had just helped to his feet. The sailor, who had come to the below-deck, makeshift clinic to be healed, was struck in the hip by the cannonball. The velocity of the projectile threw him across the small room, where he was flattened against a bulkhead. Usher Parsons rushed to his assistance, but it was too late. The impact of the projectile had ripped a gaping, two-foot hole in the sailor's back.

During the heat of battle, Parsons had shifted his surgical station from the nearly bombed-out below deck wardroom of the *Lawrence* to the deck, at extreme risk to himself. Moving from gun station to gun station across the shambles that defined the once pristine *Lawrence*, he hurried from one injury to the next, applying dressings and treating injuries. By his later count, there were almost ninety dead men sprawled in various positions on the deck. Standing upright to scan the deck, Parsons could see that the battle was finally over. The British ships were lying well downwind of his ship, the *Lawrence*, which was in shambles. *Niagara* and *Caledonia* were still in reasonably good shape, but they were not moving under sail. The smaller ships from the two fleets, six American and three British, seemed to be the least damaged, and as Parsons watched, two of the small British ships seemed to be escaping to the north. Unlike his visions of what a major naval battle should look like, longboats from both fleets were being rowed through the battle scene and moving faster than any of the large ships. Spars were lying across the decks of the ships, sails were shredded and large sections of decking were missing.

Parsons was barely twenty-five years old and terribly sick. This young surgeon's mate who was alone on the deck of the heavily damaged *Lawrence* was the only one of the American fleet's three doctors who was able to function. The combination of lake fever, seasickness and fear had incapacitated the other two American physicians. This was not quite what Usher Parsons had imagined when he signed on as a United States Navy surgeon's mate in 1812.

Parsons was not born to privilege like most physicians of his era. He was the youngest of nine children in a struggling, working-class family in rural Maine. Educational opportunities were few and far between when he was growing up, but his life of independence in New England helped him aspire to be both a patriot and an educated man. When an opportunity presented itself, he found employment as an apprentice to a doctor. Medical apprenticeships were the gateway to the medical profession for less-than-privileged boys. Children of New England wealth attended universities and earned their medical educations through academic studies before being granted the prestige of an MD degree. Undaunted by his financial need to learn medicine the practical way, Parsons assisted his hometown mentor, Dr. Abiel Hall, and studied privately. Then, in early 1812, he sat for and passed the Massachusetts medical examinations and became a licensed physician.

Equipped with a license to practice medicine, Parsons followed the authoritarian call of his personality and signed up for duty in the War of 1812's United States Navy. From a military perspective, he was the perfect candidate: a rare combination of soldier-want-to-be and licensed surgeon. Unlike university-trained physicians of his era, he had also earned the practical experience required of a naval surgeon's mate during his apprenticeship.

While Parsons, like most naval officers, longed for ocean duty and the opportunity to see the world, those assignments were generally reserved for young officers who came from upper-class New England families.

Surgeon of the Lakes

THE DIARY OF DR. USHER PARSONS
1812–1814

Edited by John C. Fredriksen, Ph.D.

Usher Parsons as a young man, depicted on the cover of John Fredriksen's book. *Courtesy of Erie County Historical Society.*

Newport's Oliver Perry, for example, received his first command on the Atlantic Ocean in 1807 after already making his mark in New England as a twenty-two-year-old naval officer. But unlike Usher Parsons, Perry grew up in a blueblood Rhode Island family as the son of a naval officer. Three years Parsons' senior, Perry seemed to represent everything that the new ship's surgeon loved about naval service. When word reached Parsons that his hero, Oliver Perry, had run his first ship, the *Revenge*, aground off Newport, was court-martialed and placed on administrative leave, the new surgeon's mate was disappointed. Even though Perry was acquitted and the fault for the loss of the *Revenge* was attributed to a civilian pilot, sinking a ship was a terrible career blow for a young officer.

Parsons took his first naval assignment in New York City in July 1812 but almost immediately volunteered for Great Lakes duty, where opportunities for promotion were greater. He arrived at the Niagara River in the late fall of 1812 in time to help treat the wounded from several land engagements. Given his newly demonstrated battlefield skills, Parsons was soon placed in charge of the naval hospital at Black Rock. The Black Rock base was an important strategic location upriver of Buffalo and behind Squaw Island, where it was protected from the British side of the river.

Parsons was pleased to take charge of the medical care of the fleet and the American soldiers who had been sent to the hospital. But he was also bored. Black Rock station was in a holding pattern during the spring of 1813, and rumors were that the next big action would be taking place at Erie, one hundred miles to the west. Prior to his arrival at Black Rock, base commander lieutenant Jesse Elliott had led a daring, stealth raid on the British naval base just across the Niagara River. Elliott and his men surprised the British and were able to seize the brigs *Caledonia* and *Adams* with the loss of only one man. In capturing the British ships, Elliott also liberated twelve American prisoners and returned the *Caledonia* to Black Rock with $200,000 worth of furs (the *Adams* went aground and was abandoned).

Inspired by evening campfire stories of Elliott's bravery and convinced that Fort Erie, where the ships had been anchored, was no longer heavily guarded, Usher Parsons organized an American raiding party using locals who were thinking of becoming a militia force. This was an opportunity for the soldier component of Usher Parsons' personality to become engaged. Parsons' first purchase when he had arrived at Black Rock was an accurate musket. He was an avid hunter who prided himself on being an exceptional shot. Armed with his personal weapon, Parsons, like Elliott,

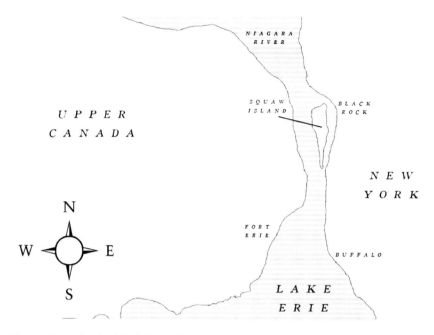

Niagara River showing Black Rock, Fort Erie and Squaw Island. *Illustration by Isaac Smith.*

led an invading party across the Niagara River on May 28 and stormed Fort Erie, which was inhabited by two unarmed civilians at the time. Dr. Parsons remains the only American surgeon to have secured the surrender of an enemy fort.

But those were among Usher Parsons' best days in the navy, and this day on western Lake Erie was hopefully his worst. It was only 3:30 p.m., and he had already performed four amputations. With as many as four assistants holding terror-stricken and shrieking patients down, Parsons had used his crude surgeon's saw to remove three legs and an arm. And he was not done. His daydreaming recollections of better days were interrupted quite suddenly when a bloody sailor with a right arm dangling loosely by a strand of cartilage fell into Parsons' arms, knocking him to the deck. By the time Usher Parsons had freed himself from the grip of the man who was at least fifty pounds heavier, the bloody arm had fallen away and it was apparent that the sailor was dead.

Chapter 2

BEFORE THE WARS

Native North Americans Meet the Europeans

*Lieutenant went onshore after wood and was fired upon by Indians. One sailor
and three soldiers wounded. Boat returned immediately.*
—Usher Parsons' diary

For almost one thousand years before European arrival, native people had inhabited the Great Lakes region. The Iroquois people, who were organized into a powerful five-nation political alliance, inhabited the southern shores of Lakes Erie and Ontario, while the Huron and Ojibwa (Chippewa) controlled the northern regions, including the Georgian Bay area. The five Iroquois tribes, ranging from west to east, were the Seneca, Cayuga, Onondaga, Oneida and Mohawk, and their political alliance, which reached from Lake Erie past Lake Ontario and into the St. Lawrence River Valley, made them a daunting force when the Europeans arrived. A number of other tribes, including the Dakota/Sioux, Winnebago and Menominee, inhabited the western Great Lakes area. Prior to the arrival of the French explorers in 1615, Indian people had established major meeting or terminal points at today's Detroit, Chicago, Toronto, Mackinac and Milwaukee.

The arrival of Europeans was perplexing and troubling to native North American people. Those who didn't succumb to the diseases that the French and English brought to the New World—frightening and incurable maladies including measles, smallpox and the common cold— quickly developed a love-hate relationship with the new white people. They were fascinated with European technology. Guns and gunpowder promised hunting efficiency and the ability to defend against enemies. And desperate as they were for directions, food and other advice, the Europeans were generous in their proclivity to trade items made of iron, including guns, traps, needles, hatchets and axes, for Indian assistance. Pots, pans, jewelry, blankets and other domestic vestiges of European culture were also popular trading items. The Indians who survived European sicknesses and made friends with the new whites benefited from their trades.

While Europeans originally traveled into the frontier using waterways such as the Great Lakes and the major rivers, Indians lived a more inland existence, carving trails through forests and using creeks to float canoes. Even though Indian people generally lived and traveled within much smaller areas than European explorers, they fully understood the geography of their lands. By the time of French arrival in 1615, for example, the Iroquois were trading with native North Americans as far south as the Gulf of Mexico. To do so, they had developed transportation pathways linking Lake Erie to the Gulf of Mexico using rivers, creeks and connector trails.

North America's first two European wars, the French and Indian War and the American Revolutionary War, were confusing and disruptive to the Indians. From their perspective, there was little difference between the French and English and even less distinction between the British and the new Americans. Indians were forced to take sides, and in many cases, the tribes or tribal factions that allied with the French against the British, and then the British against the Americans, found themselves on the "wrong political side." For the Iroquois who were divided in their allegiances, the Chief Joseph Brant faction was forced to leave its ancestral home in upstate New York and move to Ontario, where they were granted land (that had previously belonged to the Huron people) at Brantford.

The War of 1812 was even more perplexing and damaging to Indian people. Regardless of the British or American alliances that they had formed, prior decisions were suddenly called into question. More importantly, the trading relationships that had fueled the changing economy of many frontier Indian tribes were placed at risk.

British colonial interests in the New World revolved around the rich resources that were so readily available. In the northern Lake Huron and Georgian Bay areas in particular, the most valuable resource was furs. The British soon found that their best allies in harvesting the valuable furs were the Indian people who had already been using them for generations. Trading companies like the Hudson Bay Company and the North West

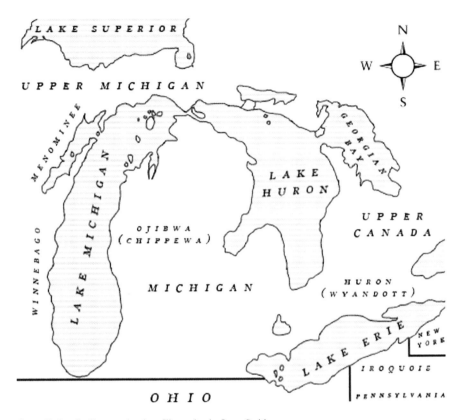

Great Lakes Indian territories. *Illustration by Isaac Smith.*

Trading Company set up trading posts on the frontier and usually found that the most secure locations for their operations were near British forts.

To maximize the return on their military investments, the British set up a separate, military-style bureaucracy to deal with North America's Indians. Within this organization, they created a hierarchy of military types known as Indian agents. The role of the Indian agent was to broker relationships between the native North Americans who lived in the Great Lakes region and the trading posts that were being used to broker beaver, fox, rabbit and other furs for Europe. As they became trusted allies of the Indians whom they represented, the agents also became advocates for Indian concerns.

After the American Revolution, British Indian agents who had been operating in Ohio, Pennsylvania, New York, Michigan and Wisconsin were forced to move north into Upper Canada (Ontario), where they were regarded by the new United States government as "provocateurs" and enemies of the state. When the War of 1812 erupted, Indian agents were identified as enemy military, even though their role in trade was primarily that of free-market wheelers and dealers.

One of the most lampooned of all Indian agents was a Pennsylvania man named Simon Girty, who had been captured by the French during the French and Indian War and given as a slave to the Iroquois. The Seneca (Iroquois) essentially absorbed Girty into their culture, where he learned their ways and their language. When the Chief Brant faction of the Iroquois moved to Ontario after the Revolutionary War, Girty became one of England's most talented and revered Indian agents. His understanding of Indian culture and language endeared him to the Iroquois, Huron and Ojibwa, and those connections helped him to orchestrate powerful trading relationships.

Girty's ability to gather intelligence and disseminate it to British forces during the War of 1812 also branded him a traitor and spy among Americans. His ability to organize the Indians and to use their skills to watch and report on American movements helped British commanders develop battle tactics and strategies. The Indians who had allied with the British did not want a disruption in the lucrative trading relationships that they had developed. In fact, Indian objection to American incursions was so strong that, as in the previous two conflicts, many took up arms and joined the British side.

Lieutenant Robert Livingston, who was attached to the Mackinac region as chief Indian agent, served in the same role as Girty. It was Livingston who was to provide British forces with the critical information that would help them in the Lake Huron and Georgian Bay actions during the war. Livingston was also able to rally thousands of Indians to join the British in their defense of Upper Canada.

Chapter 3

VICTORY ON LAKE ERIE

Limping Home

Of Dr. Parsons, surgeon's mate, I cannot say too much.
—Oliver Hazard Perry

With Perry's expanded, fifteen-ship fleet anchored in western Lake Erie, an exhausted Usher Parsons continued his grim work. For thirty-six hours after the battle, he had carefully converted the *Lawrence* into a fleet hospital by rallying all of the American medical supplies from the other boats and clearing away below-deck infirmary and surgical space. Instead of continuing with amputations the evening after the battle, Parsons chose to spend his little remaining energy making the wounded sailors feel comfortable instead of subjecting them to the terrible screams that would have accompanied more amputations. But the next morning, he was refreshed and sprung back into action.

While Perry was moving all of the ships into the shelter behind West Sister Island, Parsons used the *Lawrence*'s longboat to shepherd wounded British sailors to his ship where he treated them without regard to their enemy status. He also managed to organize the two British physicians, Drs. Kennedy and Young, and began to utilize their skills in treating all the wounded. In addition to the *Lawrence*, the British ships *Detroit* and *Queen Charlotte* were also converted into hospital ships. By the time the dead and wounded were sorted out, the total number of wounded sailors in the two fleets was 189: 96 British and 93 American sailors.

At midday on September 11, Parsons continued with the amputations. When all was done, he reported that he had amputated a total of twelve

Lithograph of the Battle of Lake Erie funeral scene. *Jerry Skrypzak Collection.*

arms and legs. Parsons even performed an emergency brain surgery to remove shards of wood and bone by drilling into a sailor's skull with a hand awl. On September 12, with his patients stabilized, Parsons helped preside at the onshore funeral of the officers. British and American officers were lined up in alternating order by country, and military music was provided by the combined American and British bands that had played during the battle. With a funeral dirge in the background, Perry read a prayer, and the officers were buried.

Perry's work was just beginning. After patching up the damaged vessels and jury-rigging *Lawrence* so that it could manage a downwind and down-current voyage back to Erie, Perry set about clearing British influence from western Lake Erie by taking Fort Detroit and Amherstburg. Perry landed 2,500 soldiers at Amherstburg where they took the British outpost without a struggle. That army marched inland to the Thames River, where it engaged British troops and won handily in a battle that claimed the life of Tecumseh. Chief Tecumseh was a fabled Indian fighter who had dogged the Americans for decades after forging an alliance with the British.

Perry also took soldiers to Detroit to reclaim the American fort there and return it to United States command. Once Perry had liberated Detroit and captured Amherstburg, Lake Erie was entirely under the control of the United States Navy. Meanwhile, Parsons and the *Lawrence* waited impatiently from September 15 to 21 for fair weather and sustained

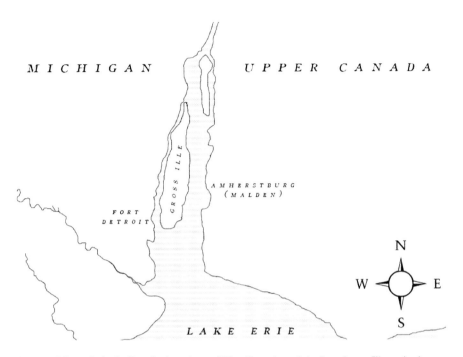

Straits of Detroit, including the locations of Fort Detroit and Amherstburg. *Illustration by Isaac Smith.*

Contemporary photograph of Put-in-Bay. *Photograph by Jerry Skrypzak.*

Perry's Monument at Put-in-Bay.
Photograph by Jerry Skrypzak.

westerly winds. It would have benefited the overcrowded group of sick and wounded from the three hospital ships to return to Erie as soon as possible. At the Misery Bay base, shore resources could have helped with patient care, but Usher Parsons resisted the urge to leave prematurely and subject his sickly patients to a gruesome passage and seasickness. Finally, the morning of September 21 brought a high-pressure system with moderate winds from the southwest, and *Lawrence* departed. Parsons and his crowded hospital ship arrived at Misery Bay two days later, following a smooth passage.

After a seventeen-gun salute and a joyous welcome at Erie, Parsons resumed his work with the wounded and sick. The badly wounded were moved onshore to the hospital building that had been set up there, and even though Parsons himself continued to suffer from lake fever, he worked relentlessly. Long after the battle had ended, it was noted that only four of Parsons' patients had died after his treatments, a remarkable success rate given the conditions under which he had worked.

It was later hypothesized that Parsons' practice of heating his surgical instruments (saws) in a bucket of boiling water had been responsible for this great success.

In reflecting upon the casualties it was probably fortuitous that both the American and British fleets were significantly undermanned. While both Perry and Barclay had pleaded with their superiors to send more sailors before the battle, the additional crew would almost certainly have created a medical burden that Parsons could not have coped with.

Chapter 4
LAKE ERIE

The Iroquois River Sea

We sailed from Erie with crews consisting of a few seamen but mostly made up of Pennsylvania militiamen. The upwind motion of the vessel alone was sufficient to disable these.
—Usher Parsons on the up-lake trip to Put-in-Bay

Even though they had lived near the shores of Lake Erie for centuries, native North Americans on both the American and British shores had generally avoided it. One frightening aspect of the lake for the Indians was the dynamic nature of the water levels, which could rise or fall significantly, especially during storms. While Indians on both shores visited the lake episodically for fishing, they generally did not settle close to the shores. When Indians did use the lake for transportation, they traveled on huge, lightweight canoes designed to carry trading goods or war parties. They were careful to remain close to shore, where they could beach their canoes in case of bad weather.

The generations of Indians who had traveled on Lake Erie concluded that it was a hybrid body of water: part sea and part river. In Iroquois language, Lake Erie was often referred to as a river-sea meaning that while it was huge, it was also characterized by a relentless river-like current. Interestingly, the Indians fully understood the size of Lake Erie and the locations of the other four Great Lakes, which they accurately described to the earliest French explorers.

Lake Erie is geographically situated with its 240-mile longitudinal axis running in the same direction as the prevailing southwesterly winds.

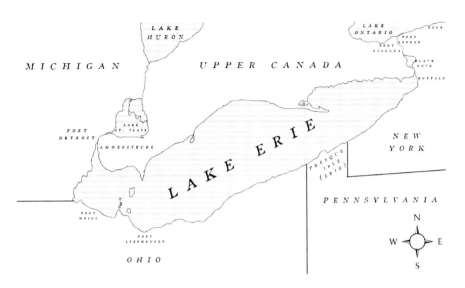

Lake Erie and its forts. *Illustration by Isaac Smith.*

Consequently, moving upwind and up-current, from east to west, was a daunting challenge for early nineteenth-century sailors. The few merchant traders of the day designed ships with shallow drafts so that they could tack into the shallows on their way up or down the lake. They also were careful to use hull shapes and sailing rigs that would allow their ships to point relatively high into the wind. Even with these advantages, travel on the lake was terribly difficult, especially for merchants who were trying to move west toward Detroit and then up the rivers leading to Lake Huron and the Georgian Bay. Adding to the difficulty of moving upwind, fall and winter storms created quick, steep waves, especially when a northeasterly wind blew against the prevailing current. The early mariners often laid up or anchored during high winds rather than fighting against Lake Erie's currents and waves.

By the time the War of 1812 erupted, there were a number of British and American forts on Lake Erie. The British held forts at Amherstburg and Fort Erie near the ends of the lake and at the northern end of the Niagara River, on Lake Ontario, where they built Fort George. The Americans occupied forts at Detroit, Toledo (Fort Meigs), Sandusky (Fort Stephenson), Erie (Presque Isle), Buffalo and Black Rock, and opposite Fort George, where they built Fort Niagara. While much of the provisioning of the British and American forts was done from land, there was an ongoing need to connect the forts using military ships.

Military records of the era indicate the extreme difficulty experienced by naval officers who attempted to make these journeys. Their frustration was compounded when ships were required to carry large numbers of troops. Regular army troops or militia who were crowded aboard ships to travel between the forts were not used to the ship's motion and often compounded the misery of long Lake Erie voyages by becoming violently seasick. As several naval officers noted in their journals, once the vomiting began in such close quarters, it was just a matter of time until everyone was miserable. Even today's mariners, professional sailors in thousand-foot bulk ships, complain about Lake Erie, with its short choppy waves and propensity for sudden and violent storms. Mariners still refer to the trip west along the axis of Lake Erie as traveling "up-bound."

When military officers described War of 1812 life on the "frontier" at Erie, they were not complimentary. Naval officer Robert Reid, upon transfer to Erie from the Atlantic, wrote, "I have never liked, nor will I ever live in this vile and wicked place. It is a collection of the filth and wickedness of both the U.S. and Canada." Lake Erie was a remote spot on the western frontier, and the town of Erie itself had fewer than fifty dwellings at the time of the war. Even at that relatively small size, Erie was the biggest town on the lakeshore. Usher Parsons, upon visiting Cleveland in 1814, wrote, "Cleveland is a small village called a city with about twelve dwelling houses. Almost every house is a tavern."

One of the ongoing problems for the navy on Lake Erie was delivery of supplies. Tools, gunpowder, cannonballs and iron parts including nails had to be supplied by overland travel from Pittsburgh or Philadelphia. One of the most heroic deliveries of 1813 was a wagon train that trundled across the wilds of central Pennsylvania to deliver critically needed gunpowder.

Chapter 5

MISERY BAY MISERIES

Lake Fever

I am reduced to a skeleton and will never cross another lake again.
—Usher Parsons, from a letter to his father

Historians who wrote about the harsh winter of 1813–14 at the United States Naval Base in Erie were shocked by descriptions of conditions there. The winter was, indeed, difficult, and the four-boat fleet—*Lawrence*, *Niagara*, *Lady Prevost* and *Caledonia*—as well as the hospital and blockhouse buildings that had been constructed onshore, continued to be overcrowded with both American sailors and British prisoners. The ships were in poor repair, and the men who had been charged with preparing them for the next year's campaign drifted between depression, fatigue and illness. Many who had been injured in the battle struggled to heal while fighting sickness. Consequently, little progress was made in repairing the fleet. Conditions at Little Bay during the winter of 1813–14 persuaded historians to rename Perry's anchorage "Misery Bay."

Usher Parsons joined his fellow fleet surgeons in describing the primary health issue on Lake Erie as a malady called "lake fever," so called because it seemed to be a sailor's illness that was only connected to service on the Great Lakes. The writings of both American and British sailors are replete with discussions of this mysterious disease. The ongoing ravages of lake fever were so difficult that Perry's scheduled departure from Erie to the western end of the lake in preparation for the September 1813 battle had to be delayed. For weeks, there were simply not enough able-bodied sailors to man the ships. Mysteriously, lake fever seemed to be at its worst when the ships were anchored or moored, especially at Erie in the protected inlet that had been named Little Bay by cartographers. On the trip west, which was both upwind and up-current along Lake Erie's axis, problems of lake fever were exacerbated by seasickness. The hastily built ships of Perry's fleet did not perform well upwind and had to be tacked repeatedly to make progress to the west. To make matters even worse, the motion of the military ships was extremely difficult and resulted in many cases of seasickness.

When ocean sailors first moved to the Great Lakes, they generally underestimated the size and power of the waters. They had been led to believe that inland sailing would be done on tiny ponds, from which they would be able to see enemy forts at all times by simply looking across the water. Their shock in discovering the immensity of the lakes and feeling the power of the waves, however, was balanced by their delight at realizing that they would be working in fresh water. Their welcomed new non-saltwater existence instantly eliminated most of the banes of a sailor's existence: nasty salt residue on everything that would not allow fabric to dry, constant concern about the rationing of drinking water, drinking dirty fluid that had been stored in warm below-deck holding

Early chart of Erie's Presque Isle Bay showing the improved government channel, circa 1900. *Jerry Skrypzak Collection.*

tanks and scurvy. Sailors who arrived at the lakes found ships floating on lovely, crystal-clear water that was far superior to the well water people had been drinking in upscale New England homes. Rather than being restricted from drinking or being issued small quantities of nasty, vile fluid from below-deck tanks, Great Lakes sailors could drink all they wanted whenever they wished.

But there was a problem. It was common practice on large ships to cut one or more holes through the foredecks in forward locations, near the head or bow stem. Sailors or soldiers who needed to relieve themselves went to one of these cutouts and "did their business." The forward position of the cutouts inspired the common name for a ship's bathroom, the "head." The reason for selecting that particular location was that once a ship began to move, residual sewage would be washed away by bow waves. Perry's ships were similarly equipped, but unlike oceangoing ships, where drinking water was stored below decks and rationed, when water was needed it was drawn directly from the lake with buckets and ropes. It is amazing that fleet physicians failed to completely connect lake fever, which is now recognized as a form of typhoid, with the relatively short distance between the place where crew went to the bathroom and the locations from which they drew drinking water.

There were clues. Lake fever was at its worst when ships were anchored instead of moving. At anchor, ships would naturally orient themselves with the bow to weather so that fecal material from the bow drifted toward the stern, where water was drawn. Lake fever was also a more difficult problem for the enlisted men than the officers. This was probably related to the fact that the ship's officers were often away from the boats on a regular basis, traveling or being entertained on shore. During the allegedly disastrous 1813–14 winter of misery, lake fever abated when the winter ice formed and sailors were forced to trudge considerable distances from the ships to cut holes through the ice, through which they fetched drinking and cooking water.

The misery of Misery Bay may also have been over-reported by historians, who made judgments about conditions from the relatively insular physical environments of their academic offices. While things were indeed difficult at the naval base, conditions were similarly harsh everywhere on the frontier. While Usher Parsons buried twelve from the compound at Little Bay during the winter of 1813–14, disease and death also continued among the civilian population in town. Two of Daniel Dobbins' children who were living at his home in the city died, as did the

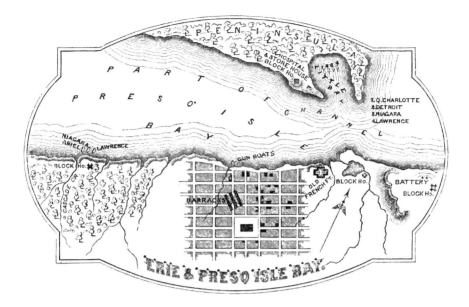

Chart of Little (Misery) Bay from 1814, depicting the town of Erie and the locations of the hospital building and blockhouse at Presque Isle adjacent to the fleet. *Jerry Skrypzak Collection.*

eldest child of Dr. John Wallace, Parsons' civilian medical colleague who consulted regularly at the naval base.

It should not come as a surprise that Parsons was largely unable to turn the tide of infection and disease that winter. He was operating in an age that preceded scientific medicine and the evolution of antibiotics. Parsons wrote about the three standard medical strategies for treating patients: (1) cathartics for stimulating diarrhea, thus purging poisons from the stomach and bowels; (2) emetics for stimulating vomiting to purge the upper digestive system; and (3) the use of needles for bleeding patients to drain toxins from their circulatory systems. The art of medicine at the time revolved around deciding which of these three strategies to use or to combine in each particular case. While Parsons would later be lauded for his innovative work in stemming infections by boiling surgical instruments, it would be decades and a return to academic medical school at Harvard before he convinced himself that the serendipitous boiling process that he used at the battle was related to his amazing rate of surgical success after the battle.

As difficult as the Misery Bay winter was, Parsons seemed to enjoy many aspects of shipboard life on America's western frontier. His diary contains evidence that he read voraciously, practiced the violin and made trips to town, where he "tarried" with new friends. Entries describe lovely teas with the town's gentlemen and ladies and the evolution of a professional relationship with Dr. John Wallace. On a number of occasions, Parsons invited town friends to join him onboard one of the ships for tea or meals. Most importantly, Parsons managed to use the winter season to heal himself from the ravages of the previous year and from the lake fever that had debilitated him before, during and after the battle. In one illustrative entry, he describes an overnight horseback hiatus to the town of Waterford:

> *Sunday November 21: A pleasant day. Went to Waterford. Seven miles south, the land is so elevated that the lake can be extensively seen. After four miles, snow was visible on the ground. Waterford is at the head of French Creek navigation and boats there readily pass to Pittsburgh and New Orleans. There are four public houses, one respectable. Passed the evening merrily.*

One important outcome of Usher Parsons' winter at Misery (Little) Bay was his growing affection for writing. Recognizing that the difference between apprentice-trained physicians and the academically educated doctors that he emulated was the ability to publish papers in journals, he began to use his free time to write. His diary and letters illustrate a growing focus on new strategies for medicine. And sometimes, as Parsons noted, the weather was not as awful as might otherwise have been assumed: "Friday February 18: Attended a show with Dr. Wallace. Weather very warm, perhaps 70 thermometer degrees at noon."

In retrospective analysis of various writings on lake fever, it is possible that the mixing of fecal material with drinking and cooking water may have been exacerbated by other factors. Parsons and others noted that crowding, poor nutrition (usually associated with scurvy) and exposure probably added to the problems of sickness at the naval base that year.

Chapter 6

NORTH COUNTRY

Sinclair's Challenge

Arrived at Lake Huron where the rapids are ¾ of a mile long and run at 4 to 5 knots.
—Usher Parsons' diary, July 1814

The Lake Huron and Georgian Bay region, north of Lake Erie, was strategically critical to American Great Lakes naval efforts during the War of 1812. Lake Huron is the second largest of the Great Lakes, with a surface area of 23,000 square miles. Measured longitudinally from south to north, Lake Huron is 209 miles long and 183 miles wide. With a surface area of 5,800 square miles, the Georgian Bay is so immense that many geographers have referred to it as the sixth Great Lake. But its gross size does not reveal the importance of the immensely rich area surrounding its shores.

The French were the first Europeans to arrive in this part of North America. Samuel de Champlain reached the southern Georgian Bay by traveling from the St. Lawrence River via the French River and Lake Nipissing in 1613. The French explorers were overwhelmed by the richness of the resources, especially the region's furs. The eastern shore of the Georgian Bay extends almost 150 miles from the southern end of the bay near today's Collingwood to Manitoulin Island in the north. Beyond Manitoulin Island, the rock-encrusted eastern shoreline continues for another 100 miles north beyond the Georgian Bay boundary, past Drummond Island and St. Joseph Island, where its inside waters rejoin Lake

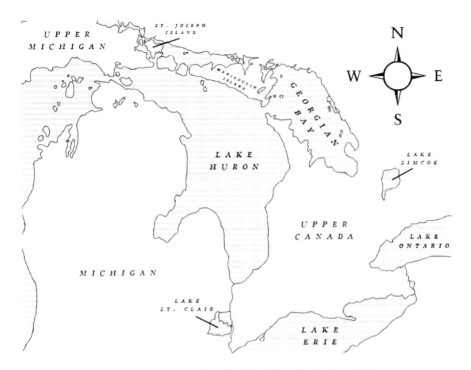

North Country: Lake Huron and the Georgian Bay. *Illustration by Isaac Smith.*

Huron. The rugged eastern shoreline provided a perfect environment for beaver, mink, ermine and other valuable fur-bearing animals. Geologically, the eastern shore, which was carved by the glaciers, has been identified as part of the Canadian Shield. It is composed of granite bedrock, which has distributed itself into the more than thirty thousand islands that populate its coastline.

After the French and Indian War, the governor general of Upper Canada (Ontario) set up a naval base at the southern end of the Georgian Bay at Penetanguishene on Matchedash Bay, the site of an earlier Ojibwa village. Jesuit missionaries had established mission trading posts in the area by the late 1620s as French voyageurs began traveling regularly between Montreal and the Georgian Bay.

The northern end of Lake Huron, which joins with the upper waters of the Georgian Bay, had long been one of the most strategically important locations on the Great Lakes. It was at this juncture that Lake Huron (and the inside waters north of the Georgian Bay) was joined to both Lake Superior and Lake Michigan. The French River and Lake Nipissing connection was the traditional French transportation route from the east. This passage terminated in the Georgian Bay, just east of Manitoulin Island. From this point, the voyageurs would canoe and portage along the protected waters behind Manitoulin and Drummond Island to St. Joseph Island and the fort that was located there.

At the northern end of Lake Huron, there were two important trading routes to the other Great Lakes. The Straits of Mackinac joined Lake Huron with Lake Michigan at Mackinac Island, and the St. Mary's River flowed into Lake Superior at St. Joseph Island. Two important island fortresses guarded the entries to Lake Michigan and Lake Superior. St. Joseph Island and its Fort St. Joseph was situated at the Huron end of the St. Mary's River, and Mackinac Island with its fort(s) sat at the entrance to the Straits of Mackinac, at the entry to Lake Michigan. For the French and then the British, these two islands were critical terminal points for the fur trade. Both

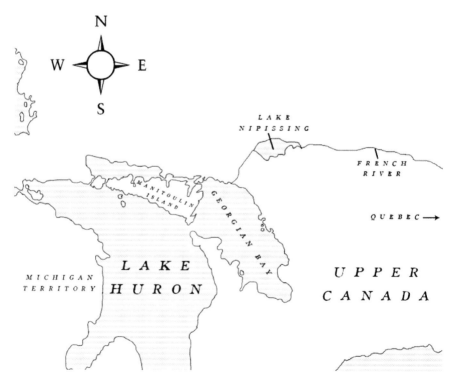

The French River and Lake Nipissing passage to the Georgian Bay. *Illustration by Isaac Smith.*

had long been established meeting places for native North Americans prior to European arrival. For the Indians, however, they were meeting points for tribes who lived within reasonably short distances of the islands. Unlike the Europeans, who were used to moving huge distances, beginning with the trek across the Atlantic to the New World, Indian trade had been a chain of short- and medium-distance travels, with individual tribes usually moving a few hundred miles or less to make contact with each other.

The French built their first fort on the mainland at the Straits of Mackinac and called it Michilimackinac. After the French and Indian War, the British took over the old fort but quickly decided that it was not in the best strategic location. The British decided that a better location would be across the straits on Mackinac Island itself. They built a fortified structure on Mackinac, using the island's natural limestone cliffs to augment the structural fortifications that they created. After the American Revolutionary War in 1796, the British reluctantly turned their fort over to the Americans. The British left Mackinac and moved to St. Joseph Island, where they built Fort St. Joseph, which was on the British (Upper Canada) side of the international border. Only sixteen years later, in July 1812, a combined force of British regulars and Indians from Fort St. Joseph, in ten bateaux and seventy war canoes, retook Mackinac. Commanding officer Robert McDouall defiantly raised the British flag once more, but this time within American (Michigan) territory. This act was seen as a major insult by the Americans.

With the success of the 1813 Battle of Lake Erie, and Fort Detroit in the hands of the Americans once more, the United States Naval Administration was anxious to take control of Lake Huron and the Georgian Bay and to interrupt the lucrative fur-trading business that British Indian agents had been using to finance the war. With complete control of Lake Erie and a formidable fleet of warships, the only major obstacle was the treacherous passage from Lake Erie to the North Country. Arthur Sinclair was selected as the new commander of the Lake Erie fleet, and all that stood in his way was the preparation of the Erie fleet and then the daunting trip upriver through the Straits of Detroit, Lake St. Clair and the Straits of St. Clair, a distance of approximately eighty-four miles. Sinclair was an acclaimed naval sailing master with the rank of master commandant, but nothing from his ocean sailing experiences against the British could have prepared him for the challenges of North America's frontier North Country.

The first northern Indian traders and voyageurs to arrive at Lake Erie called attention to the riches of the Lake Huron and Georgian Bay region by paddling huge canoes packed with furs down the straits to Detroit, a feat that

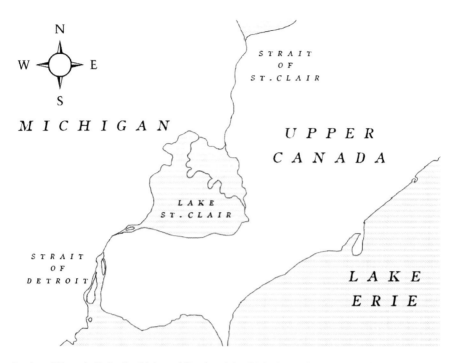

Straits of Detroit, Lake St. Clair and Straits of St. Clair. *Illustration by Isaac Smith.*

was made easier by the swift currents that characterized the two river systems. No one noticed the ways in which these traders made their way back up the straits, but the typical technique was to tow their empty canoes back upstream from shore and to portage around the most difficult sections of river. In many portions of the straits, the current was simply too strong to be negotiated.

Lake Huron and the Georgian Bay lie at 577 feet above sea level, almost 10 feet higher than Lake Erie. While it would seem that such a short gravitational fall distributed over eighty-four miles would not be daunting, nothing could be further from the truth. The currents and bends in both the Straits of Detroit and St. Clair made upriver sailing nearly impossible. The earliest traders who negotiated this challenging waterway did so with great difficulty and by using every trick, including anchoring, kedging and towing from shore. They also used relatively small ships, like Daniel Dobbins' Erie-based *Salina*, which was a relatively shallow-draft vessel at six feet and was capable of pointing high on the wind. Sinclair's challenge was to move his behemoth military ships with their deep keels up the waterway against the current.

Chapter 7

DANIEL DOBBINS

Sailing Master

Based upon Perry's description of the captured British fleet, a prize of
$225,000 was awarded.
—United States Congressional Committee report

Daniel Dobbins was born in 1776 near Lewistown, Pennsylvania. After the state of Pennsylvania purchased the triangle of land adjacent to Lake Erie, he moved to Erie with Judah Colt to become a Great Lakes sailor and helped to build and launch the merchant sailing schooner *Lady Washington*. Their ship, which was launched in 1796 at Freeport just east of Erie, was the first American commercial sailing vessel on the upper Great Lakes. In the early 1800s, Dobbins left Colt and went to work for the Reed family in Erie. Dobbins became their senior captain with an owner's stake in the schooner *Charlotte*, which he renamed *Salina*. At the onset of the war, he was clearly the most knowledgeable sailor on the upper Great Lakes.

In 1812, Dobbins was at Mackinac Island trading salt for furs when the British raided the American fort there. His schooner was impounded, and he was sent as a prisoner to Amherstburg by way of Detroit, which had also fallen into British hands. The British had offered amnesty to anyone at Mackinac who would pledge never to take up arms against England, but interestingly, Dobbins refused and was sentenced to death. Dobbins managed to escape and immediately traveled to Washington, D.C. There, he met with Secretary of the Navy Paul Hamilton to brief him on the strength and locations of the British navy on Lake Huron and Lake Erie.

While he was there, he managed to convince Hamilton that the best place to build a new Lake Erie fleet would be at Erie, Pennsylvania, where the peninsula and its channel with a sandbar guarding the harbor would prevent the British from sailing by and blasting the under-construction ships to bits. Dobbins further fueled the case for Erie as a new naval base by describing the availability of oak trees near shore for laying keels and softwood trees for planking.

When Dobbins left Washington, he was given a sailing master's commission in the United States Navy, a budget of $2,000 and a mandate to immediately begin production of four gunboats. When Jesse Elliott at Black Rock heard that Dobbins had secured the Erie location, he was furious. Elliott argued that the fleet should have been constructed at Black Rock, where it could have been better protected from the British. To attack a Black Rock fleet, the British would have had to sail down the Niagara River from Fort Erie, where they would have been out of navigational control in the extreme current and easy targets for land-based cannons. Elliott also noted that from the Black Rock base, an American fleet could have been launched directly up the Niagara River, even though Dobbins argued that oxen would have been required to tow the ships up the fast-running water to the eastern end of Lake Erie.

When Perry arrived in Erie for a preliminary February 1813 visit of the new shipyard, he wasn't sure what to make of the curmudgeon who was in charge. Dobbins was only thirty-six at the time, but given Perry's youth (he was only twenty-seven) and inexperience with Lake Erie, the sailing master found Perry's intrusion more than a bit irritating. Perry, in turn, was disturbed by the fact that the yard workers were calling Dobbins "captain," a title that he had acquired through merchant experience rather than military appointment. In the navy, the rank of captain was a step higher than commodore. The naval shipbuilders whom Perry brought with him were not pleased with the dimensions of Dobbins' in-progress gunboats and immediately ordered him to add ten feet to their lengths, a decision that Dobbins never agreed with. Ocean sailors with absolutely no understanding of the sailing characteristics of Lakes Erie and Huron, or the treacherous Detroit and St. Clair River currents, were telling the most respected Great Lakes sailing master how to design ships.

When Perry returned to Erie later to take up permanent residence at the base, he brought shipbuilder Noah Brown with him from the Atlantic. Both Brown and Perry continued to hold Dobbins in some suspicion, even though the local sailing master had consistently solved every ongoing construction problem, including the lack of iron. It was Dobbins who showed Brown how to fashion wooden ship nails from hardwood and use them in place of traditional iron fastenings.

As if to emphasize their suspicions of Dobbins, Perry and Brown opened a second shipyard west of the original downtown site, which from their perspective may have been a move that was made necessary by the deeper water there. It was at the Cascade Creek Yard where they built the two large ships *Lawrence* and *Niagara* and the *Ariel*. After Perry's ten-ship fleet had finally been moved over the sandbar and into the lake using camels (floating dry docks used to lift deep-draft vessels) for *Niagara*

Daniel Dobbins. *Jerry Skrypzak Collection.*

and *Lawrence* (Dobbins claimed to have understood and presented the lifting system in Washington as a reason for using Erie as a site), Perry assigned Dobbins to the schooner *Ohio*. Interestingly, in all the reports that Perry subsequently sent to the navy, the size of the American fleet was listed at nine, not ten, ships. The *Ohio* was never included. It seemed strange to some of the sailors that the one person with extensive Lake Erie sailing experience would not have been aboard Perry's flagship, the *Lawrence*, for the Battle of Lake Erie.

Instead, Dobbins was relegated to supply duties with the *Ohio*, even though he claimed that his armed ex-merchant ship was a more capable sailing vessel and pointed higher on the wind than any of the others. At the time of the actual battle, Dobbins was aboard the *Ohio* moving between Erie and Put-in-Bay with supplies, even though he was listed on the crew manifest as sailing master aboard the *Lawrence*. A second critical mission for Dobbins and the armed *Ohio* as it moved Perry's supplies, was to prevent any goods from reaching Robert H. Barclay's British fleet at Amherstburg. When the battle was over, Dobbins asked why the American fleet had even bothered to engage the British, since they were desperate for supplies and starving to death at Amherstburg.

Perry concluded that Dobbins did not deserve a portion of the $225,000 in prize money and took Dr. Usher Parsons into his confidence to explain why he had decided not to share the prize money with Dobbins. At the same time, Perry confided with Parsons that although Jesse Elliott had failed the fleet by not following the *Lawrence* into the fire at the beginning of the battle, he had decided to award Elliott and his crew with their proper share. Parsons appreciated this confidence and wrote, "I have impressed the Commodore [Perry] with such a favorable opinion of me that I have not the least doubt that he will render [career] assistance to me for a better situation. He is the first warm friend that I have made in this [naval] service who is capable of assisting me. In the course of the year I hope to obtain a little prize money."

Dobbins was infuriated by Perry's attempt to deprive himself and his crew of their share of the prize money. Although he continued in the navy for decades, he never spoke positively about the celebrated naval commodore, even though his negativity proved harmful to his own career. Some have speculated that Perry was trying to protect Dobbins, since the shipwright was technically a wanted man. If the British had captured Dobbins at a battle, he would almost certainly have been executed. If this was true, it may have been a generous act by Perry to attempt to keep the fleet shipwright from the battle.

Chapter 8

THE BURNINGS

From York and Dover Mills to Washington, D.C.

The fleet sailed last evening for long point transporting militia, 500 regulars
about 400 with the purpose of burning public property.
—Usher Parsons' diary, May 14, 1814

One of the American practices that most infuriated the British was the burning of public property. Destruction of military facilities, forts and blockhouses was understandable, but when United States troops set fire to civilian buildings, British authorities who were trying to win the trust of Upper Canada's population became outraged. This was particularly true since so many of the settlers along the northern shores of the Great Lakes were United Empire Loyalists (UELs) who had moved there after fleeing America during the time of the Revolutionary War.

The first significant War of 1812 burning was conducted by American troops from Sackets Harbor, New York, on Lake Ontario. Under the direction of Isaac Chauncey, Jesse Elliott and others sailed to York (Toronto) on fourteen ships, landed more than 1,500 troops and attacked the fort and its naval docks. Attempts to defend the city by several Indians under the command of a British Indian agent and a few British regulars were unsuccessful, and within a few hours on the morning of April 27, 1813, York was defenseless.

A handful of citizens and a small detachment of local militia attempted to surrender, but the Americans were angered when a British military ship at the dock exploded. British regulars had lit a fuse leading to the ship's

Lithograph of York (Toronto) shown in the early 1800s. *Ian Bell Collection.*

gunpowder as they fled. The ship, which would have been an important war prize, exploded as the Americans approached it on the docks. The soldiers were already calculating the prize money that would have been theirs if they had returned it to the United States side of the lake. Anger over the explosion riled the landing troops, who retaliated by looting and setting fire to private homes and businesses.

In Erie, the spring of 1814 brought changes to the Misery Bay base. Jesse Elliott departed for the Atlantic Ocean, Dr. Usher Parsons was promoted to fleet surgeon based on the recommendation of Perry and, in April, a new base commander, Arthur Sinclair, arrived. When Sinclair inspected the base and its ships, what he found did not please him. In fact, he wrote a scathing report criticizing Elliott for the terrible condition of the ships, which he had hoped to launch immediately. Sinclair wrote: "The materials have not even arrived for repairing them. They [the ships] are in the same shattered condition as when they came out of the battle." Elliott later defended his lack of progress, suggesting that Sinclair did not fully understand the difficulty of moving shipbuilding materials from Pittsburgh or Philadelphia or the sickness and malaise of the men. Dobbins later added that little progress was made because he had been assigned to spend the winter in downtown Erie with his supply ship, the *Ohio.*

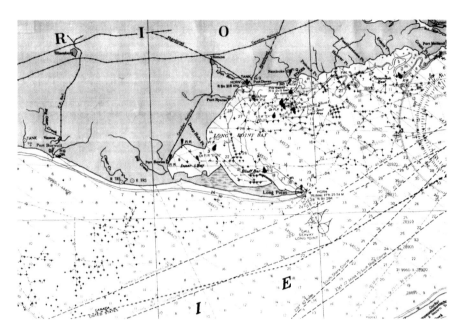

Chart of Long Point and Long Point Bay showing Port Dover, Port Ryerse and Turkey Point. *Jerry Skrypzak Collection.*

Sinclair began by transferring Parsons' hospital operations from the *Lawrence*, which he judged to be almost ready to sail, to the captured British *Detroit*. The focus of his repairs was on *Lawrence* and *Caledonia*, which he prepared and moved over the bar, taking advantage of early spring's high water levels. In May, Lieutenant Colonel John Campbell arrived with troops in anticipation of a burning raid of the Long Point region. The May action was, in effect, to be a warm-up for the 1814 Lake Huron Campaign that was to follow. While Sinclair remained in Erie to work on his ships, five hundred regulars and militia were loaded aboard *Lawrence* and *Caledonia* and sailed across the lake to the north shore, where intelligence suggested that there were three active grain mills. The Dover Mills, Port Ryerse and Port Rowan gristmills were helping supply the British and Campbell's mission was to disrupt their operations.

Campbell's troops arrived at Port Dover behind Long Point on May 14. The next day, his troops rowed to the beach in longboats, marched into the village of Dover Mills, where they met almost no resistance and burned the entire town, including its gristmill, after looting food and supplies. The next morning, the ships moved west to Port Ryerse, where troops burned the

Today's Public Pier at Port Dover, Ontario, has become an important town center. *Photograph by D. Greenwood, provided by Tracy Haskett and Norfolk County Tourism and Economic Development.*

Contemporary photograph of Port Ryerse, Ontario, the town where Barclay attended the 1813 party and was then burned in the 1814 raids by Campbell. *Photograph by Earl Hartlen, provided by Tracy Haskett and Norfolk County Tourism and Economic Development.*

Contemporary photograph of the Backus (Backhouse) Mill, the only Long Point-area gristmill to survive Campbell's 1814 burning raids. *Photograph by Jerry Skrypzak.*

Ryerse farm, a sawmill and another gristmill. Later that day, the ships sailed for the Inner Bay and Port Rowan, where Campbell planned to burn the Backhouse saw- and gristmills. Head winds made the sail difficult, however, and Campbell feared that tacking into the Inner Bay and passing Turkey Point, where there was known to be a British fort, could subject his troops to musket fire from shore. After several hours, Campbell ordered the sailing crew to ease the yards and head back toward Erie. Campbell's decision to give up his plan to burn the Backhouse mills may also have been influenced by the fact that his troops were out of food. When Sinclair learned that Campbell's troops had burned private dwellings, he issued a letter of outrage, calling the conduct disgraceful.

American burnings continued through the 1814 Campaign. In fact, Sinclair's own troops later burned public property on the Georgian Bay at St. Joseph Island. In August 1814, the British landed a major force on the Chesapeake Bay and marched to Washington, D.C. Claiming retribution for the outrageous burnings of public property at York and Dover, British regulars set fire to the American Capitol building and houses of Congress as President Madison and his staff fled.

The United States government later issued an official apology for the burning of Dover Mills and sent a financial settlement check. Interestingly, Dover Mills officials decided to move the town east and rename it Port Dover, thus laying the groundwork for the thriving tourist and commercial fishing center that it has become. The Backhouse (Backus) Mill survived and has become the centerpiece of an Ontario Heritage Park in Port Rowan, where much of the original wooden mill still stands.

Chapter 9

MILLER WORSLEY'S
FIRST COMMAND

The Georgian Bay Supply Line

*The water is colder than I ever saw any spring water. It cannot be held in the
mouth or used to clean teeth until warmed.*
—*Usher Parsons' diary entry while sailing up Lake Huron*

British control of Lake Huron and the Georgian Bay began when
American troops at Mackinac Island were overwhelmed in 1812. Their
presence on the strategic island at the confluence of Lake Huron and Lake
Michigan placed the British in control of the western lakes. But a strategic
problem emerged for the British in early 1814: supplying the remote island
outpost. American military actions had succeeded in stopping or slowing the
traditional overland supply lines, and the British were desperate to develop a
new way of moving food to the remote post.

Summer comes late on Lake Huron, so it wasn't until July that newly-
promoted Miller Worsley, a twenty-three-year-old first lieutenant, was
assigned to command the Georgian Bay end of Mackinac Island's critical
supply line. Worsley was a British regular, having joined the navy from his
home at Isle of Wight in 1803. In 1805, based on his experience at the Battle
of Trafalgar, he was promoted to midshipman.

Worsley was wiry and thin, but muscular. Onboard the ships that he had
served on in the Atlantic, he had earned a reputation for toughness. He
was also comfortable in the wilderness and seemed to get along well with
all kinds of people, from Ojibwa Indians to the voyageurs and regular navy

Southern Georgian Bay, including Matchedash Bay, Penetanguishene, Notttewasaga Bay and the Nottewasaga River. *Illustration by Isaac Smith.*

men sometimes assigned to his command. By early July, Worsley was at his new Georgian Bay post with his men, organizing his small fleet of ships and staging supplies.

The British had decided to move their Georgian Bay supply base from Penetanguishene at Matchedash Bay to the mouth of the Nottawasaga River, fewer than ten miles south. Their decision was based on ease of material movement, and the trek was pioneered by Lieutenant Colonel Robert McDouall, Worsley's immediate superior, who was in charge of Mackinac. Worsley's supply fleet was composed of a pair of one-masted rowing bateaux that had been constructed by carpenters at Mackinac and an eighty-foot supply schooner named the *Nancy*. Worsley's sailing ship was a commercial vessel launched in 1789 for trade on the western lakes. Designed for the big waters of Lakes Huron, Michigan and Superior, it was stout by comparison with military ships and built to carry a huge 350-barrel payload. At the onset of the war, the *Nancy* was requisitioned for military service and named HMS *Nancy*.

In addition to his schooner and bateaux, Worsley's post included his second in command, a regular navy midshipman, twenty-one sailors, nine French Canadian voyageurs and twenty-three Ojibwa Indians. The first delivery step at the outpost involved moving supplies from York to Nottawasaga Bay. Supplies were moved in barrels from York via Yong Street, which was an ancient Indian trail, to Barrie. At Barrie, Worsley's troops portaged the supplies to the Nottawasaga River, where the barrels were loaded onto the bateaux and towed down the Nottawasaga to its mouth at the southern end of the Georgian Bay. The length of the land-water route was about 80 miles. After staging at Nottawasaga Bay, barrels of supplies were loaded aboard the *Nancy* and sailed north, up the Georgian Bay and then out into Lake Huron to Mackinac Island, a distance of 250 miles.

By the first week of July, Worsley had already delivered two loads of supplies to Mackinac. Upon unloading the first supplies at the northern island outpost, he was stunned to see the deplorable conditions of the troops. After a long, hard winter, supplies had reached a critical level, and Worsley realized that he had to move as rapidly as possible to return with more food. It was usually a two-day sail from Mackinac to Nottawasaga Bay, and as he guided the *Nancy* south after his second trip, he mulled over rumors that an American fleet would soon be in Lake Huron trying to disrupt his work. As soon as he arrived at Nottawasaga, one of Worsley's Ojibwa seemed anxious to speak with him. The scout quickly informed Worsley that the Americans were on their way and struggling up the rivers in sailing ships.

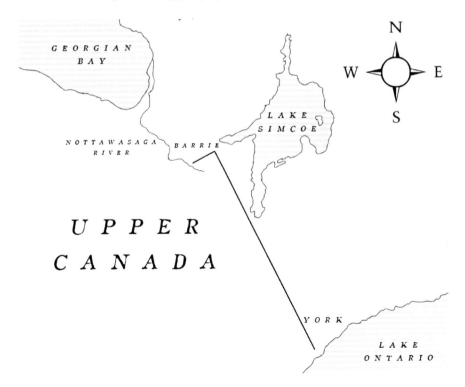

Miller Worsley's York to Barrie to Nottewasaga Bay Supply Line. *Illustration by Isaac Smith.*

Alarmed that the Americans were coming so soon, Worsley immediately departed with a handful of Ojibwa and his second in command. Always a man of action, he decided on a scouting mission to see if he could spot the American fleet. It was a long hike across Indian trails from Nottawasaga to the Lake Huron shoreline, and when Worsley's party finally approached the beaches, several Ojibwa were already there to meet them. The Ojibwa began an excited conversation with Worsley's Indian confederates, pointing toward the open waters of Lake Huron.

It was the morning of July 16, and as Miller Worsley and his party crawled on their bellies toward the beaches of Lake Huron, eight sailing vessels slogged along in light winds with their sails luffing badly. Worsley was taken aback to see that three of the biggest ships, brigs by his calculation, were crowded with troops. Hundreds of American troops were huddled on the decks, looking like they had never before been aboard a ship. On the *Lawrence*, a few miles offshore, Usher Parsons was busy writing in his diary: "The commodore concluded that he had been wrong about making for Matchedash Bay and he set course for Michilimackinac [Mackinac] instead. The wind is nearly on our larboard [port] bow."

The reality of this major American incursion sent Worsley into action. He immediately dispatched several Ojibwa to Mackinac to warn McDouall. Then he departed, running and walking, for his base at Nottawasaga. When he arrived, Worsley and his men immediately went to work hiding the barrels of supplies that had been staged for the next shipment and fortifying the blockhouse that was being constructed upriver on the Nottawasaga. Worsley could not be sure if the Americans were coming for his base immediately or sailing directly for Mackinac.

When the American fleet did not arrive during the next few days, Worsley decided that he should try to hide the *Nancy*. Working feverishly, his troops unloaded the ship, removing cannons, munitions and other supplies. Once the *Nancy* was lightened, they towed it upriver for almost two miles into shallow water. On August 10, Lieutenant Robert Livingston of the Indian Department arrived in his thirty-five-foot canoe. Livingston had departed Mackinac and paddled to Nottawasaga Bay with additional Ojibwa to warn Worsley that the Americans were on their way.

After hiding Livingston's canoe, the men at the Nottawasaga Base removed *Nancy*'s yards and sails to make the schooner blend into the forest backdrop behind it. Once they had accomplished that task, they dug in to wait and hope that their small detachment would not be discovered. All the while, Worsley was obsessing over the supplies that were so desperately needed at Mackinac.

Chapter 10

DESTINATION MACKINAC

The 1814 Campaign

On the bar, Scorpion *and* Tigress *came alongside and lightened us some but when released, we drifted to shore where we spent the night.*
—*Usher Parsons' diary, July 5, 1814*

Still seething over the deplorable condition of the fleet and embarrassed by the actions of his ships at Dover Mills, Arthur Sinclair left the Erie base with a five-ship fleet on June 18, 1814, bound for North Country. Sinclair was aboard *Niagara*, and his flagship was joined by the *Lawrence*, *Caledonia*, *Tigress* and *Scorpion*. Aboard the *Lawrence*, Usher Parsons' responsibility was to prepare his ship hospital for battle, as well as any injuries or sicknesses that might occur en route. Before he departed, Parsons turned the care of his Erie patients aboard the *Detroit* and at the onshore hospital building to his civilian colleague, Dr. John Wallace.

Sinclair's mission was to retake the former American fort at Mackinac Island that had embarrassingly been surrendered to the British without a struggle, and also to find and destroy or capture the British supply ship *Nancy*. The loss of Mackinac Island had been a terrible embarrassment, and the continuing British presence there was a black eye for the American military. Since the northern overland supply lines to Mackinac had been cut off, Sinclair realized that conditions at the British outpost would be desperate. That was why he had hoped to leave on his mission earlier, before the British could use the *Nancy* to deliver food and supplies. He was also aware of the difficulty of the trip up the Detroit and St. Clair Rivers, which is why it seemed strange to Daniel Dobbins that he was not asked to join the expedition.

En route to Lake Huron, Sinclair stopped at Detroit to load the military troops that had arrived from Sandusky under the direction of George Croghan. Lieutenant Colonel Croghan's force included 750 soldiers. Of these, 400 were regulars, and the rest were Ohio Volunteers. Sinclair's fleet left Detroit on July 3, at least two weeks late by his calculation. In addition to the five ships that had departed Erie, four smaller sloops from Detroit joined Sinclair to assist with the ascent of the rivers and help with supplies. The presence of the troops aboard *Niagara*, *Lawrence* and *Caledonia* slowed progress and complicated the

Arthur Sinclair. *Courtesy of Erie County Historical Society.*

operation of the sailing rigs aboard the large ships, which were forced to execute countless tacking maneuvers. The restless troops also demanded more food than Sinclair had anticipated.

There were two primary navigational problems on the arduous upriver trip to Lake Huron. The first was that the large ships had extreme difficulty negotiating the fast-running river currents. The prevailing winds were essentially from the west, which gave the ships some sailing advantages, but given the river bends and currents, *Niagara*, *Lawrence* and *Caledonia* had difficulty making progress. The smaller ships were used to assist by towing. On several occasions, longboats were launched for kedging or so that rowers could help with propulsion. The second and more difficult navigation issue, however, was Lake St. Clair. While it was seemingly large and passable, there were many shallows and sandbars. Without an experienced pilot aboard, the fleet took unfavorable tacks and repeatedly ran aground, especially the deep-draft *Lawrence* and *Niagara*. The smaller ships were used to lighten *Niagara* and *Lawrence* when they went aground, ferrying troops to shore to wait and removing the heaviest of the stores.

Sinclair finally had a bit of luck when he reached the end of the St. Clair River, where the current and rapids were ordinarily the most treacherous. The wind freshened, and on July 13, he was able to sail into Lake Huron,

George Croghan. *The Frick Art Reference Library*.

where he anchored and waited to reassemble his troops. On July 14, Sinclair was finally ready to sail north. Returning one of the supply ships to Detroit, he headed north for the Georgian Bay entrance just beyond Tobermory, where he planned to sail south for Penetanguishene and Matchedash Bay to look for the *Nancy*.

The winds were cooperative, and by July 16, the fleet was abeam the area where they had expected to find the entrance to the Georgian Bay, but they could not locate it in the haze. By this time, Sinclair could see that he had a new problem: his supplies were in jeopardy. One unanticipated issue in Lake Huron was the poor fishing. The crew had been able to catch large quantities of delightfully mild-tasting fish in Lake Erie, but on Huron, where the water was much deeper, they were having little success, and food stores were disappearing rapidly. Sinclair ordered the remaining provisioning ships to offload enough food to the big ships so that one of the small ships could return to Detroit and he wouldn't have to feed its crew.

On July 20, Sinclair reached St. Joseph Island, where he expected to find British regulars at Fort St. Joseph. When his troops were sent to shore, however, they were disappointed to learn that the former British fort had been abandoned. Sinclair ordered his men to burn it and return with any food or supplies that they could find. While they were on the island, his men also burned the abandoned headquarters of the British North West Trading Company and slaughtered all the cattle they could catch. On July 21, a strange sail was spotted on the horizon. Assuming that it was the *Nancy*, the gunboats chased it down, and it surrendered. The ship turned out to be a British commercial schooner, the *Mink*. Sailors from the *Mink* told Sinclair that the British Georgian Bay supply center was not at Matchedash but that it had moved to Nottawasaga Bay.

On July 28, with the crew on half rations, Sinclair finally reached Mackinac. Hoping that a well-planned attack would quickly take the fort

Northern Lake Huron, with Mackinac Island and St Joseph Island. *Illustration by Isaac Smith.*

and allow access to any food that the British had in storage, Sinclair and Croghan spent days reconnoitering and crafting a battle plan. After several scouting trips along the shores of the island, however, both officers were shocked to see the extent of the British fortifications. Sinclair later described the fort as being "impenetrable as Gibraltar with a rise of 200 feet to perpendicular rock cliffs on three sides." They first tried an artillery assault from the ships, but the cannonballs fell harmlessly short because the fort was so high on the island's bluffs. Sinclair finally decided that the fort was out of range. With supplies continuing to dwindle, they launched a troop landing at the same beach where the British had disembarked when they took the island two years earlier. Using cover fire from the ships to clear the way, Croghan's troops landed on the low ground at the beach and marched uphill toward the fort.

The battle did not go well. British commander Robert McDouall outmaneuvered the American troops, which had been divided, with the Ohio volunteers taking a flanking action while the regulars marched directly uphill toward the fort. McDouall used the Indians who were at his fort to engage the Ohio Militia and his regulars against the advancing American regulars. Within a few hours, all of the Americans had been repelled and were retreating to the beach. The immediate toll was twelve killed, including several officers, and forty wounded. Several of those died later as Usher

Photograph of today's Mackinac Island with the fort shown high above the modern tourist city. The fort's elevation prevented Sinclair from reaching it with *Niagara*'s cannons when he couldn't raise their firing angle sufficiently. *Photograph by Jerry Skrypzak.*

Mackinac Island, the Straits of Mackinac and the Upper Michigan Peninsula mainland, where Fort Michilimackinac was originally built. *Illustration by Isaac Smith.*

Parsons was ministering to them aboard the *Lawrence*. The next day, a party carrying a white flag went ashore to ask for the body of the slain major who had led the charge. McDouall met the Americans, politely offered the body with his apologies and provided the landing party with food and cooking fuel, even though his own supplies were desperately low.

On August 6, a disappointed Arthur Sinclair weighed anchor and sailed south. The trip back down the lake and rivers would be much easier, with fair winds and advantaged currents but this was little consolation. In spite of the British generosity in providing supplies for Sinclair's trip home, Sinclair decided that he should still try to intercept and capture the *Nancy*, thus accomplishing at least one facet of his mission.

At the entrance to the Georgian Bay, Sinclair dispatched the *Lawrence* with Usher Parsons and the wounded to Erie. *Caledonia* and the smaller ships, including the captured *Mink*, were to accompany the *Lawrence* as far as Detroit.

Chapter II

THE *NANCY*

A Consolation Prize

*A sail was discovered N by E and apparently 15 or 20 miles. The vessel was
probably the* Nancy.
—*Usher Parsons' diary, July 16, 1814*

At dawn on August 13, 1814, Arthur Sinclair eased *Niagara*'s yards and
headed into the Georgian Bay. Convinced that he would find the
Nancy sailing inside the protected waters of the bay, he ordered *Scorpion*
and *Tigress* to fan out on either side of his *Niagara* as the three ships reached
to the south. Sinclair's convoy carved a swath that was more than ten miles
wide as his ships headed south toward Nottawasaga Bay. Given their sailing
angle, it would have been impossible for the *Nancy* to avoid detection if it
was in the bay.

But the *Nancy* was nowhere to be seen, and Sinclair's three-ship squadron
slowly closed on the southern shore of Georgian Bay without finding their
prey. As evening fell, the American ships anchored to wait for the British
supply ship to arrive. Meanwhile, Miller Worsley and his men were dug
into the sand fewer than two miles from the American ships, watching.
Worsley was joined in his subterfuge by Lieutenant Robert Livingston and
his Indian companions.

Convinced that they might be waiting for some time, Croghan led his
soldiers ashore to set up a camp and relieve the crowding aboard *Niagara*.
A few hours after they had disembarked, a party of Croghan's soldiers,
foraging for firewood spotted the *Nancy* in its upriver hiding place. Croghan

The British supply ship *Nancy. David Brunelle, Laurel Finney and Wasaga Beach Provincial Park.*

quickly reported his find to Sinclair, who, in turn, moved *Niagara* as close as he could to the camouflaged supply ship and anchored. Then Sinclair began to fire with the *Niagara*'s cannons. But the *Nancy* was out of range, and cannonballs were falling harmlessly on the sand, well short of their objective. Unsure of the size of Worsley's force, Sinclair offloaded several smaller pieces of artillery (carronades) and had Croghan's men move slowly toward the *Nancy*, alternating musket fire with the artillery.

At the *Nancy*, Worsley could see the hopelessness of his situation. As the first wave of American troops advanced, three of his men were hit. One was killed and two wounded. Not wanting the Americans to capture the *Nancy*, even though it was empty, Worsley placed explosive charges in both the ship and the blockhouse, which was only a few hundred feet away. Then, during a lull in artillery fire as the Americans moved their cannons, Worsley lit the long fuse that led in both directions and fled. He and his men moved upriver, towing the bateaux and Livingston's canoe with them. Back at the beach, a second round of artillery fire seemed to result in a major explosion and fire aboard the *Nancy*, which ultimately burned to the waterline. Croghan reported that he had hit the British ship with a cannonball, but Worsley later said that it had exploded and burned because he had set a fuse that ignited gunpowder aboard the ship.

While Worsley's men regrouped upriver, Sinclair was celebrating his victory and breaking into the barrels of food that he and Croghan found near the British blockhouse. The next morning, Sinclair left in the resupplied *Niagara* for Detroit, instructing the two smaller ships to remain together near the entrance to Nottawasaga Bay in case trading ships carrying furs or supplies sailed within sight. As *Niagara* sailed away, Worsley's Ojibwa friends watched from shore. The Indians continued to observe the remaining gunboats for several days until the captains aboard *Scorpion* and *Tigress* decided that they should move north, where they would be more likely to encounter fur traders or other shipping that was bound for Mackinac.

As *Scorpion* and *Tigress* departed, Worsley was returning to the beach to take stock of his situation. The *Nancy* was destroyed, but he still had the two bateaux, as well as Lieutenant Livingston's large canoe. Worsley was also heartened to find that the Americans had missed one hundred barrels of the supplies that he and his men had hidden.

Chapter 12

SNEAK ATTACK

Miller Worsley's Georgian Bay Victory

My birthday, very pleasant. The Porcupine [from Erie] *is in sight with letters no doubt for the Commodore* [Sinclair]. *Troops sailed for Detroit in small vessels.*
—*Usher Parsons' diary, August 18, 1814*

On August 18, a few days after the American gunboats *Tigress* and *Scorpion* sailed from Nottawasaga Bay, Miller Worsley sprung into action. Based on Indian agent Robert Livingston's report, he realized that McDouall and the troops at Mackinac would be dangerously low on supplies. So he loaded most of the one hundred barrels of food that he had recovered into his two bateaux and began to paddle north with Livingston and his canoe as an escort. For as dangerous and difficult as this trip seemed, none of Worsley's crew questioned the mission. Twenty-five men, two bateaux and an oversized canoe were about to attempt to paddle the length of the Georgian Bay, some 230 miles, enter Lake Huron and then make a dangerous 10-mile trip through open water to Mackinac.

Six days into their incredible journey, Worsley's heart sank when his overloaded and ponderous bateaux rounded a cape on Drummond Island and spotted the two American gunboats sailing toward them. Reversing course, he quickly guided his fleet to shore, where he and his men dragged their boats into an inlet and covered them with branches. An hour later, *Tigress* and *Scorpion* both sailed by. Fortunately for Worsley, the Americans were focused in the opposite direction, away from shore, and had not seen

Contemporary reenactor portrays Miller Worsley. *David Brunelle, Laurel Finney and Wasaga Beach Provincial Park.*

them. Worsley realized what they were doing aboard the American gunboats: watching for canoes that might be loaded with valuable furs. For American captains, capturing trading canoes and taking the cargo could pay great financial dividends. The furs would be counted as a war prize, for which the captains and crews would be rewarded a proportional share.

Leaving a few men to guard the bateaux and supplies, Worsley crammed eighteen of his sailors with Livingston into the thirty-five-foot canoe at dusk on August 25 and began to paddle north again. During the middle of the evening, he managed to sneak past both American gunboats using an old Ojibwa trick: stretched animal skins fastened to the paddles to muffle sound. At dawn on September 1, Worsley arrived at Mackinac. After a hurried greeting, Worsley met with McDouall and presented a daring plan. McDouall had doubts, but recognizing Worsley's determination and his own desperation, he agreed to provide his young lieutenant with two gun-mounted bateaux and a detachment of sixty Newfoundland fencibles (marines). Anxious to put his plan into action, Worsley departed that evening just after dark, hoping to take advantage of diminishing winds for his open-water return trek across upper Lake Huron.

Compared to the loaded bateaux that he and his men had been rowing, the war bateaux were easy to propel. After stopping once to rest and eat, Worsley spotted the *Tigress* by itself, anchored off the northern end of Drummond Island. Just as he had anticipated, the American gunboats had separated, hoping to improve their chances of spotting trading canoes. Quietly rowing the two bateaux under cover of darkness until one was on either side of *Tigress*, Worsley gave a signal, and a brief firefight erupted. Worsley's sailors and the Newfoundland marines then leapt aboard *Tigress*

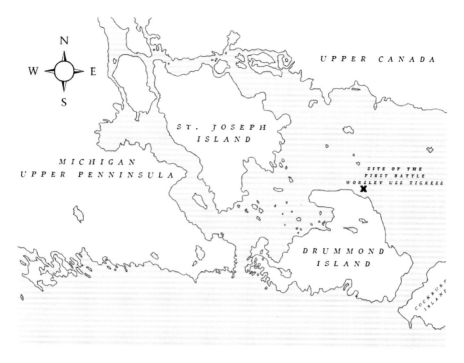

St. Joseph Island and Drummond Island showing the site of the first battle between Worsley and the *Tigress*. *Illustration by Isaac Smith.*

brandishing swords. The Americans were caught off guard, and when the brief conflict was over, Worsley's ragtag crew had captured the American gunship. There were three casualties on each side, and Worsley's troops took the remaining twelve American sailors prisoner.

The next day, Worsley held a funeral ceremony on shore and buried the dead. Then, on September 4, he took the American prisoners to shore, provided them with food and supplies and left them on Drummond Island to fend for themselves. Then he returned to *Tigress*, raised the American colors and went sailing. Using information that he had obtained from the captured crew, he expected to find *Scorpion* five or ten miles to the south. On September 6, Worsley spotted *Scorpion*, hailed it from afar and headed toward the ship. The bulk of his crew were hidden below in their British uniforms. When he closed in on the unsuspecting ship, his crew unleashed a volley of musket fire, and the troops poured onto the deck. The fight was brief, and by midday, Worsley was in command of both American gunships. Two Americans were killed in the second fight, but no British.

On September 10, Worsley proudly returned to Mackinac in the two former American ships with the bateaux in tow, including the two that he had hidden earlier on Drummond Island with the barrels of supplies. McDouall greeted Worsley's triumphant return with a full display of ceremonial colors and a rousing brace of military music. The two American gunboats were renamed HMS *Confiance* (*Scorpion*) and HMS *Surprise* (*Tigress*). *Surprise* was so named for the manner of its capture. After resting for a few days, Miller Worsley left Mackinac to resume his supply duties, using his new two-ship "British" supply fleet.

Chapter 13
FLEET SHIPWRIGHT

Curmudgeon, Logistics Specialist and Spy

The Battle of Lake Erie was a desperate contest fought between the hungry and the sick.
—*Walter Rybka, captain of the brig* Niagara

No other person on the frontier, including the Indians who lived near the shores of Lake Erie, knew as much about the territory as Daniel Dobbins. Even though he had such detailed knowledge of the territories and was the acclaimed sailing master of both Lake Erie and Lake Huron, his role at the Presque Isle (Misery Bay) Naval Base had systematically evolved into the fleet shipwright, a position that he imagined to be a waste of his knowledge and skill. Before he entered the navy in 1812, Dobbins had already spent sixteen years tacking up and down Lake Erie and into both Lake Huron and the Georgian Bay, where he traded cargos of agricultural products, iron, salt and furs. Dobbins had served as the captain of several merchant trading schooners, first for Judah Colt and then for the Reed Family of Erie, sailing into every creek-side inlet on the north and south shores of the lake.

During his trading days, Dobbins took schooners into such out-of-the-way British places as Dover Mills, Port Ryerse, Turkey Point, Port Rowan, Port Stanley and Leamington to discharge and take on cargo. He had friends and associates in every corner of Lake Erie, both American and British. Many of his British contacts on the north shore were United Empire Loyalists, former settlers from the American side of the lake who fled to Upper Canada after the War of Independence. As a result of his experiences and associations, he, more than anyone else on Perry's 1813 or Sinclair's 1814 fleets, knew the shapes and contours of every ship landing on Lake Erie. More importantly, he had contacts at the villages along both the north and south shores.

Intelligence gathering was made somewhat easier on the early 1800s frontier by virtue of the fact that almost everyone who was involved on either side of the British/American conflict was of white, European descent. While there were a number of African American sailors on the American side, and Indian participants on both sides, the officers/ intelligence gatherers were predominantly European. The business of military spying began in 1813, when Barclay set up his blockade east of Erie. As Barclay's fleet lay at anchor that summer, he sent longboats to shore on a regular basis. His men changed to civilian clothing, hiked into Erie and returned with regular reports on the progress of the construction of the fleet. Dobbins, who was busy preparing the fleet for battle, was well aware of British intelligence-gathering activities and may have provided the spies with confusing information regarding progress of the fleet. That would have partially explained Barclay's cavalier departure from the blockade and return to Port Dover that summer.

One of Barclay's biggest problems at Amherstburg, in addition to feeding his sailors and trying to complete work on the *Detroit*, was the care of the almost one thousand Indians who were living near his British fort. The Indians had pledged loyalty to the British, and in return, they had been offered food and shelter. It was pressure to feed the Indians that had convinced Barclay that his best strategy was to untie his fleet and engage the Americans in September 1813. Had he abandoned the post and marched inland, he would have had to deal with the angry Indians. When the battle had ended and Barclay and his men were taken prisoner, the Indians dispersed, spreading along the north shore of Lake Erie where they became potential intelligence sources.

While Perry was well aware of Dobbins' connections to people around the lake and his ability to interpret the meanings of activities on shore, he did not take advantage of these skills by asking him to gather intelligence in 1813. There was far too much urgency to prepare the ships for battle. Instead, Perry used Dobbins as a shipwright and logistics specialist. After criticizing Dobbins' original ship designs, Perry reluctantly grew to count on his shipwright skills as the two men supervised construction of the fleet at the two Erie yards.

The most devastating thing that happened to Dobbins was the fact that while Perry was using him and his ship the *Ohio* as a supply vessel, he had missed the actual battle. Thinking that the fighting might not take place for several more weeks, Perry sent Dobbins back to Erie for a fresh supply of meat and fruits. While it is doubtful that it could have been true, Dobbins often told people that he knew the battle had begun because he heard the cannon fire from his anchorage at Erie while he was loading the *Ohio*. In addition to being disappointed that he and his ship did not participate in the actual battle, Dobbins was infuriated by the fact that Perry did not acknowledge him in the official navy reports. For years after the battle had ended and Perry had departed, Dobbins fought for his share of the prize money, which he ultimately received by petitioning directly to Congress.

Post-battle prize money was not distributed in a particularly democratic way. Perry and Elliott each received $7,100, a handsome sum in those days. The other captains' shares were each $2,300. While Dobbins was ultimately successful in securing a captain's share, he was terribly distraught over not receiving one of the twenty-four commemorative battle swords that were awarded to all the officers. Perry's official rationalization for not awarding a sword to Dobbins was that the shipwright had not been with the fleet during the battle. Dobbins spent more than twenty years trying to arrange to have a battle sword awarded to himself, and even though his old superior, Jesse Elliott, lobbied on his behalf, he was never successful.

Dobbins' supply ship was a welcome sight for the tired and wounded sailors from both fleets when the *Ohio* arrived at Put-in-Bay, and Perry was generous in distributing food to all the survivors, including the British. As soon as the *Ohio* was unloaded, however, Perry took it over and began using it as an escort ship and troop carrier. He needed the relatively shallow-draft *Ohio* to conduct funerals following the battle and to ferry troops to Amherstburg and Detroit. Meanwhile, Dobbins was assigned the task of repairing the wounded ships, especially the *Detroit*. After consulting with Barclay, Perry decided that the *Detroit* was probably the best of the fleet's ships, having been built meticulously at Amherstburg. So Perry had his shipwright spend extra time returning it to seaworthy condition, a task that Dobbins resented since it was a British vessel. His preference would have been to dedicate his energies to the American ships and sink the *Detroit*.

After Perry left Erie later that year, Dobbins and Elliott got along well during the winter of 1813–14. But in the spring, when Arthur Sinclair took over the American base, Dobbins was again faced with a bureaucratic blueblood as a commanding officer. Like Perry, Sinclair was a polished, eastern gentleman who spoke in an upscale New England dialect. The two men never liked each other, and Sinclair blamed Dobbins for the poor condition of the fleet when he arrived. While he knew that the crusty merchant sailor was his best hope for repairing and preparing the ships for 1814, he continued to distrust Dobbins.

When John Campbell left Erie on the first of Sinclair's ships to be prepared that spring and sailed to Long Point and Dover Mills, it was Dobbins who provided navigational information and intelligence. Dobbins had been sailing in and out of Long Point Bay for more than a decade. It was also Dobbins who warned Campbell about the potential dangers of tacking into the Inner Bay within musket range of Turkey Point. But even though he had intimate knowledge of the land forms on the British side of the lake, Dobbins was not asked to accompany the troops on the Dover Mills raid or to travel north on Sinclair's Mackinac mission later that summer. Instead, he alternated duties among repairing the ships at Misery Bay, using the *Ohio* to provision Sinclair's ships and transporting troops on Lake Erie.

Unlike Perry, however, Sinclair understood the potential for using Dobbins' knowledge of the north shore for intelligence gathering. During the late summer and fall of 1814, when military action had shifted to land, Sinclair dispatched Dobbins to the British shore to gather intelligence several times. Under the guise of trading for food and grain, Dobbins would take the *Ohio* to farming villages along the north shore of the lake. Then, while he was arranging for supplies, he would send men inland to learn of troop movements.

In 1815, when the war ended, Dobbins continued at Erie, and Sinclair was replaced by a new base commander, Captain Daniel Dexter. Dobbins was angry that he had been passed over for the job and displeased that Dexter had been sent to Erie to close the base. More loyal to his hometown than the naval officials whom he resented, Dobbins fell into disfavor with Dexter as he argued to keep the base open. To distract Dobbins in 1816, Dexter sent him off on an extended northern mission that ended in Green Bay, Wisconsin. With the base essentially closed and only the *Niagara* still in use, Dobbins' general attitude toward the navy continued to foul. In 1819, he was court-martialed for a fistfight in which one of his fellow officers was severely beaten. As a result of the court-martial, he was suspended without pay for eighteen months. The general opinion with respect to officers and their conflicts was that they should have been settled by dueling and that fistfighting was not a

Daniel Dobbins' family home near Erie's waterfront. *Erie County Historical Society, MacDonald Collection.*

gentlemanly way to settle disputes. Dobbins' proclivity for fistfighting was not befitting his officer's status and helped fuel the general notion that he was not a refined gentleman and that he was "operating out of his element." During the transitional period between 1820 and 1825, the base continued on paper, but even the *Niagara*, which had been serving as Erie's last remaining naval ship, was scuttled at Misery Bay. Dobbins was ultimately reinstated and returned to duty at Erie.

Always a thrifty man, Dobbins managed to parlay his hard-won prize money into an entrée to Erie's elite business class. He became a partner in a number of local business ventures, including Erie's newest shipbuilding enterprise, the Erie and Chautauqua Steamship Company and later the Erie to Pittsburgh Canal. While he continued to resent the military establishment that had slighted his contributions to the war effort, his naval panache made him a popular figure among citizens of the city.

While Dobbins seemed to be a hard and grizzled man, he privately felt remorse and sadness for the difficulty that his naval career had caused his wife, Mary, and his children. Beginning with the wartime deaths of his two daughters Elizabeth and Eleanor, which he blamed on his connections with the sickly men at Misery Bay, he realized that much of his career had been a terrible distraction from his loved ones and that his remaining children—especially his oldest son, Decatur, who was terribly troubled and eventually committed suicide—had suffered from his inattention.

Another new base commander, Captain David Deacon, arrived in Erie and was placed in charge. Deacon disliked Dobbins even more than the prior commanders, and as the 1820s progressed, Dobbins found himself in one administrative scrape after another. The dilemma for Captain Deacon, however, was the growing community admiration for their local hero. During the 1820s, Dobbins managed to build a mansion for his family on the bay front and become a ranking officer in the local freemasons and a founder of Erie's St. Paul's Episcopal Church. Dobbins' popularity convinced Captain Deacon that the best thing he could do would be to orchestrate a transfer for Dobbins to the Atlantic Ocean. Choosing family over career, Dobbins refused the transfer and resigned his commission.

Chapter 14

PERRY AND ELLIOTT

The Duel that Never Took Place

Dueling was an accepted military practice for maintaining honor until the late 1800s. The most famous American duel took place between Secretary of State Alexander Hamilton, who was fatally wounded, and Vice President Aaron Burr, at Weehawken, NJ in 1804.
—*Ryan Chamberlain*, Pistols, Politics and the Press

The war officially ended on February 17, 1815, when the United States ratified the 1814 Treaty of Ghent. But the conflict between Perry and Elliott was just heating up. While Jesse Elliott was distracted by duty in the Atlantic, his old commander was mustering political support, which was fanned by friends and family in Rhode Island. His supporters were lobbying on his behalf for a number of honors as his naval hero acclaim was rapidly spreading. One difficulty that plagued Perry supporters, however, was the festering controversy regarding stories that had been told after the battle. While Elliott had never publicly spread negative stories about his commander, Perry and his supporters had not been so circumspect. Perry's Rhode Island friends even launched a movement to withdraw the Congressional Medal of Honor that had been awarded to Elliott.

   Most troubling for the military panels that had reviewed the results of the battle and examined Perry's claims was the fact that Perry had originally commended Elliott in written documents, crediting his strategy of remaining at long-gun distance as he rushed in with the *Lawrence*. Later, however, Perry seemed to change his mind noting that his immediate

Oliver Hazard Perry. *Jerry Skrypzak Collection*.

reports were intended to save Elliott's career. Another troubling aspect of the post-battle analysis was the testimony of British commander Barclay in 1813 during his court-martial in Portsmouth, England. Barclay's court-martial was a customary procedure rather than an indictment of his bravery or incompetence. British commanders were regularly court-martialed after losses so that military tacticians could learn from their experiences. In Barclay's case, however, there were several adjunct issues that seemed troubling.

The fact that Perry won the battle did not come as a surprise to the British admiralty. Barclay was outmanned, six ships to nine, and placed in a situation at Amherstburg where his men were running dangerously low on supplies. His flagship, the *Detroit*, was fitted with substandard, leftover land cannons from the fort at Amherstburg instead of proper marine ordnance. Barclay's practical choices in September 1813 were either to starve to death at the dock or untie the ships and fight.

But there were three matters of concern with respect to Barclay's actions. First was his summer of 1813 abandonment of the blockade at Erie that would have prevented Perry's ships from leaving the harbor. A continued blockade would have prevented any possibility that the Battle of Lake Erie could have taken place long before it ever began. A second line of questioning was related to the fact that after he had abandoned the blockade, Barclay had attended a party at Port Ryerse while Perry was moving his ships out of the harbor. According to Amelia Harris Ryerse, the British commander arrived at the gathering with a "pretty young widow lady" who had been aboard his ship during the blockade. While the details of Amelia Ryerse's observations were not recorded until several years later, she said that as Barclay raised a toast, he boasted that he had returned to Long Point to move his young lady friend out of harm's way. (The young woman's presence aboard Barclay's ship was never verified.) Then during the toast, he announced that he was about to sail back to Erie, where he "expected to find the Yankee brigs hard and fast on the bar in which predicament it would be a small job to destroy them."

Robert Heriot Barclay. *Library and Archives of Canada.*

The third matter was Barclay's failure to destroy the helpless American fleet as it was being lifted over the bar when he had returned to Erie.

A seriously wounded Barclay stood before the court, propped on crutches. With a badly mauled leg and his one remaining arm shattered to the point of no longer being useful, he introduced letters of support from Perry, who commended the British officer on his bravery, and others. Perry had convinced his fellow American officers to join in his commendation of their British adversary. He suggested that Barclay's abandonment of the Erie blockade was motivated by the fact that he was out of supplies and had to return to Long Point to re-provision. And, according to Perry, the reason why Barclay had failed to attack the helpless American fleet as it lay defenseless on the bar was that he could not see from his offshore position in the haze that day that the cannons had been offloaded from the *Lawrence* and *Niagara*. The unusual east wind that day resulted in both the *Lawrence* and *Niagara* facing in the same direction, even though *Niagara* was hard aground and being lifted over the bar. Then when Perry executed a bold bluffing maneuver by sending *Ariel* and *Scorpion* toward the British fleet, Barclay decided that he was sailing into a trap. Barclay came about and used the favorable east wind to sail west for Amherstburg where he could complete the construction of the *Detroit*, which would be the largest and most powerful ship in his fleet. Given his intelligence regarding Perry's fleet, Barclay testified that he would have been significantly outgunned at Erie.

After reading Perry's letters and listening to testimony regarding the bleak situation on Lake Erie with Amherstburg's supply lines cut, the military court acquitted Barclay. Barclay did remark that he had been troubled since the battle by Perry's overly solicitous treatment of him. He noted that Perry had taken him aboard his own ship and treated him with great respect and dignity and even paid for him to stay in a hotel in Erie. But in hindsight, Barclay was troubled by Perry's generosity, and as he began to hear of Perry's latter-day allegations that Elliott had exhibited cowardice by not following *Niagara* into close combat, Barclay noted that it was Elliott's decision to lay back along with *Caledonia*, which had turned the tide of the battle in favor of the Americans.

The testimony in Portsmouth, England, did not have much of an impact on pro-Perry sentiments in the United States. After the awarding of Congressional Medals of Honor to Perry and Elliott, Perry supporters stumped their way up and down the Atlantic coast arguing for making Perry into a national hero. No one was making such an argument for Elliott, even though he had also received the same medal. Then, in 1817,

Perry fueled the situation by launching a lawsuit in which he accused Elliott of cowardice.

When Elliott returned from Europe shortly after Perry had launched the suit, he was troubled, embarrassed and shocked. Distracted as he had been, Elliott was only vaguely aware that trouble had been brewing from Perry and his political supporters. Elliott's response to Perry's allegations was typical of the warrior officer that he was: he publicly challenged Perry to a duel at the time and place of his choosing. Perry was not much of a duelist. Only once before during his career had he been faced with such a challenge that he could not defuse. In early 1817, while in Europe, Perry was challenged to a duel by a marine officer with whom he was serving. The two men had become involved in a messy conflict that stimulated a military court action. The court was reluctant to proceed since Perry was a war hero, but it did issue Perry a letter of reprimand, noting that as a superior officer, he should not have slapped the insolent marine. The military court judgment did not satisfy the angry marine officer, and the duel was arranged.

At the appointed time of the duel, Perry strode smartly away from the marine during the count and then dodged when the marine fired and missed. Perry aimed his single-shot pistol at his adversary and then refused to fire. Perry argued that his conduct at the duel was heroic. Dueling was a messy business in those days. While most duelists avoided direct-shooting deaths during the actual gunfights, infections from gunshot wounds proved fatal for a large proportion of combatants. When Elliott made his challenge, Perry refused to acknowledge it.

The Perry-Elliott dispute was a terrible embarrassment for the navy, as well as the United States government. Perry and his supporters had friends in high political places, and as major newspapers began to cover the affair, Perry's naval superiors struggled to find a way to quiet the matter. Finally, Perry was offered a South American mission as a way to separate the two men and quiet the festering controversy. In late 1818, Perry shipped off to Venezuela, silencing the controversy and helping to avoid the pressure for him to duel Elliott.

Chapter 15

DESTINY

Perry, Sinclair, Worsley, Barclay, Elliott, Dobbins and Parsons

Passed the day at Dr. Wallace's. Orders arrived for Messrs. Holdup, Breese and myself to repair to Baltimore and report ourselves to Captain Perry on board the Java. High glee.
—*Usher Parsons' diary, December 2, 1814*

Six of the seven characters whose fates were altered by their connections to the Erie fleet moved on to other endeavors. With the exception of Dobbins, each continued his naval career and then made a life in other parts of the world. The seven are presented in chronological order of their deaths.

OLIVER HAZARD PERRY (1785–1810) was the first of the six to pass away. In 1819, while on the South American cruise that the navy had arranged as a way to silence the controversy between him and Jesse Elliott, he contracted yellow fever. Perry died while his shipmates were desperately attempting to get him to a hospital in Trinidad. He was originally buried in Port-au-Prince, but his body was later interred and moved to his hometown of Providence, Rhode Island. It is ironic that the commander, who was said to have been blessed with unbelievably good luck, finally fell victim to such a disease. He survived the absolute annihilation of the *Lawrence* while most of his crew were wounded, managed to row through heavy fire from the *Lawrence* to the *Niagara* without being hit and then died of yellow fever. Perry continues to be one of the most memorialized war heroes in United States history, with dedicated statues and memorials at a number of sites, including Providence, Rhode Island; Erie, Pennsylvania; and Put-in-Bay, Ohio. Hundreds of cities, schools, parks, hospitals and other memorials have also been named for him.

ARTHUR SINCLAIR (1780–1831) left the Erie base in 1814, convinced that it was impossible to run a "proper" naval operation away from the Atlantic Ocean. His time aboard *Niagara* in 1814, in the limited "sea room" available on the Great Lakes, reinforced the fact that his acclaimed sailing skills were wasted on the inland seas. He softened his criticisms of Elliott and Dobbins after experiencing the logistical issues of life on the American frontier in 1814. Sinclair was transferred to the Atlantic in 1815 after the war and then to Washington, D.C., in 1818. In 1819, he was promoted to commodore and placed in charge of the important base at Norfolk, Virginia. In 1821, he established a sailing school at Norfolk aboard the frigate *Guerriere*. His active role at the school allowed him to teach the skills that had made him a famous sailing tactician in 1812, when he was cut off from his fleet in the mid-Atlantic and then led a six-vessel British attacking force on a three-day chase, eventually eluding them. Sinclair was acclaimed to have been the best pure sailor in the navy at the time. He passed away in Norfolk in 1831. During the Civil War, his three sons, who had been naval officers, resigned their commissions and joined the Confederate navy. His grandson Upton Sinclair became one of America's leading novelists.

MILLER WORSLEY (1791–1835) worked the British supply line between York, the Georgian Bay and Mackinac Island using the two captured American ships *Confiance* (*Scorpion*) and *Surprise* (*Tigress*) until ice formed on the northern lakes in late 1814. As he continued his work, he became progressively sicker and was finally forced to return to York, where he was diagnosed as having lake fever. In 1815, the British navy sent its Georgian Bay hero home to the Isle of Wight to live on a partial pension. He married in 1820 and had three children. In 1832, he was appointed inspector for the British Coast Guard. Worsley never fully regained his health, and in 1835, he died at the age of forty-four. Had he not passed away, he would have been awarded the new British Medal of Valor, but there were no provisions for posthumous awards of that particular medal at the time.

ROBERT HERIOT BARCLAY (1786–1837), like Perry, was young at the time of the Battle of Lake Erie. Born in Scotland, he was only twenty-seven when he was placed in command of the British fleet on Lake Erie. He was born to a military family and continued his Scottish clan's reputation for bravery when he lost his left arm in a close-quarter sword fight aboard a naval ship in 1809. After the War of 1812, he married his first cousin and remained in England trying to recover from the wounds that he had suffered on Lake Erie. While recuperating, he continued to petition the Crown for a new command and was finally granted a last posting as captain of a domestic bomb ship from 1822 to 1824. In 1825, he retired to Scotland on a military pension and lived there until he passed away in 1837 at age fifty.

JESSE ELLIOTT (1782–1845) continued his naval career, taking a number of assignments in the Atlantic. He fought in the Second Barbary War, commanding the sloop USS *Ontario* for three years. Elliott was promoted to captain in 1818 and sailed throughout South America through the 1820s. He was made commander of the West Indies Squadron from 1829 to 1832, after which he became commander of the Boston Naval Yard in 1833. In 1835, Elliott became commander of the Mediterranean Squadron and began a long administrative post in Europe. He was a legendary tough guy who refused to put up with interpersonal challenges. His gruff manner caused him political difficulties in the Mediterranean, and he was involved in a number of scuffles. His worst offense, according to his junior officers, was filling his ship with Sardinian donkeys that he planned to return to his farm in Maryland. This act deprived everyone else of the space needed to bring souvenirs home from Europe. For this offense, he was suspended for several years until President Tyler pardoned him and returned him to duty as commander of the Philadelphia Naval Yard in 1844. It was during his suspension that he authored the paper in which he carefully told his side of the story of what happened at the Battle of Lake Erie. Elliott died in 1845 at age sixty-three.

DANIEL DOBBINS (1776–1856) continued both his naval career and his life in Erie, somehow managing to remain attached to the naval base in his hometown even after the base was ostensibly closed down. Dobbins resigned his naval commission in 1826 when he was transferred to the Atlantic. He had many friends in Erie, including the politically powerful Reed family, and he used their influence to argue for the establishment of a protective revenue service in the city. In 1829, he was granted his wish and awarded command of the revenue cutter *Benjamin Rush*, which was stationed downtown just a few blocks from his home. His revenue service career was politically spotted by decommissions when federal political parties shifted. He was removed from the service in 1841 when the Whigs were elected and reinstated in 1845 under President Polk. He resigned his commission in 1847 after being assigned to Norfolk, Virginia, but was reassigned to Erie again during the Zachary Taylor administration in 1848. Dobbins resigned his revenue service commission in 1849. Dobbins was a mercurial character who counted good friends and extreme enemies among his associations and continued to tell stories about his War of 1812 experiences. Dobbins was fond of recounting the extreme difficulty that

Perry had in moving the Black Rock ships up the Niagara River when they came to Erie in early 1813. It took 200 soldiers and 50 sailors five days to haul the Black Rock fleet upstream and into Lake Erie, a task that clearly supported Dobbins' argument that Erie was a better location for a Lake Erie naval base. He had many business interests, among which was the purchase (with a partner) of the old Misery Bay naval yard and the ships that had been scuttled there. His financial acumen made him a wealthy man. Daniel Dobbins left behind three grown children when he passed away in 1856 at the age of eighty.

DR. USHER PARSONS (1788–1868) continued his naval career, serving with Perry. He left Erie in 1815 to join Perry on the *Java* and sailed with him to the Mediterranean. Parsons never again served as a battle surgeon. In 1817, obsessed with becoming a traditionally trained academic physician, he took a leave from the navy and attended Harvard Medical School, where he earned the prestigious MD degree in 1818. He returned to the navy and sailed to the Mediterranean again for two years. In 1820, Parsons resigned his commission and took a faculty position at Dartmouth. The following year he moved to Brown, where he joined the medical faculty. In 1822, while traveling in Rhode Island's elite social circles, he met and married Mary Holmes, sister of famed orator Oliver Wendell Holmes. He and his wife had one child. While continuing at Brown, he became a founding member of the Rhode Island Hospital and subsequently was elected president of the state chapter of the American Medical Association (AMA). Parsons' pension for writing served him well both socially and professionally. He continued to write about his experiences as a naval surgeon, as well as his affection for Oliver Perry. Parsons was also a prolific medical author, publishing more than fifty articles and papers during the 1850s and 1860s. Usher Parsons passed away at age eighty in 1868. In 2001, he was inducted into the Rhode Island Hall of Fame.

Photograph of Dr. Usher Parsons as a medical school professor in Rhode Island. *Jerry Skrypzak Collection.*

Chapter 16

PERRY'S FLEET

The Ravages of Time

Sinclair has no regard for the feelings of his officers. This is universally admitted.
—*Usher Parsons' diary, September 28, 1814*

As Sinclair's ships headed back toward Lake Erie after the 1814 Campaign, Usher Parsons continued to minister to the injured and sick from the failed Mackinac attack. Parsons first moved aboard the captured *Mink*, from which he towed the bodies of the slain officers back to Detroit for burial. After leaving Detroit in September, Parsons made several trips back and forth across the lake with Sinclair while his new commander provided support for troops at the eastern end of Lake Erie.

Finally, by late October, Parsons and the ships had returned to winter at Misery Bay. Sinclair prepared the fleet for winter by ordering the ships to be arranged with the lowest-profile vessels facing beam to the prevailing winds and separated by twenty to thirty feet so that moving air would prevent stagnation and reduce onboard sickness. Sinclair was planning to leave and move to the Atlantic, but he fully expected the fleet to be reactivated in 1815. Regarding his experiences with Perry's fleet, he wrote, "Bad sailing vessels filled with Kentucky troops, and every island and shore filled with savages, can you wonder if I have grown grey?"

On December 8 and 9, Parsons said farewell to his Erie friends, including Dr. John Wallace, Seth Reed and Rufus Reed. Rufus Reed presented him with a gift of twenty dollars. On December 10, Dr. Usher Parsons left Erie and the Misery Bay base to Join Oliver Perry on the *Java*.

Of the sixteen ships that had been credited to Perry after the battle, only six were at Misery Bay that winter. These included *Lawrence, Niagara, Detroit, Queen Charlotte, Porcupine* and *Caledonia*. The *Tigress* and *Scorpion* had been captured by Miller Worsley in the Georgian Bay. The *Ohio* and *Somers* were captured that summer by the British at Fort Erie and became British military supply ships. The badly damaged British *Chippewa* was cut loose and beached shortly after the battle. The other five vessels—the British *General Hunter, Lady Prevost* and *Little Belt*, and the American *Trippe* and *Ariel*—had been taken to other bases on Lake Erie. *Trippe* ran aground near Buffalo and burned to the waterline in 1814. The *Little Belt* was left at anchor near Buffalo during the winter, and British troops crossed on the ice and set it on fire. *Lady Prevost* was sold to a businessman on Lake Huron and converted to a supply ship.

When the war ended in 1815, Erie's naval base was closed and the ships that remained in Misery Bay, including *Lawrence, Detroit* and *Queen Charlotte*, were scuttled except for *Niagara*, which continued to serve as a harbor greeting ship until 1820, when it was also sunk. A Rochester businessman purchased Misery Bay's scuttled ships in 1825 from the United States Navy. He raised *Caledonia*, converted it to a commercial ship and sold it to Erie's Rufus Reed, who renamed it *General Wayne*. The *Queen Charlotte* was

View of Misery (Little) Bay from the 1920s showing Perry Monument, the *Wolverine* and houseboats. *Jerry Skrypzak Collection.*

similarly converted for merchant duty and sold to Erie's George Brown, who shortened its name to *Charlotte*. The *Detroit* was raised, converted and sold to a businessman in western Lake Erie.

Historians have noted that the War of 1812 and its fleets of military ships were a powerful demonstration of the potential for commercial shipping on the Great Lakes. As a result of this sudden awakening, there was a sudden and insatiable demand for new commercial sailing ships on Lake Erie. As ill suited as military ships were for commercial duty, this was the niche that several scuttled veteran warships filled. In 1836, a local man named George Miles, at the urging of Daniel Dobbins, purchased the remaining vessels at Misery Bay. Miles and Dobbins subsequently became partners in the remaining Misery Bay collection of scuttled ships.

In commenting on Sinclair's criticism of his command at Erie, Jesse Elliott noted that on his watch he did not lose a major battle or have four ships taken by the British. By this time, however, Sinclair was venting his anger against the captains of the *Scorpion* and *Tigress*. In a letter to his friend John Cocke, Sinclair wrote: "I left them 25 additional men, supplies and boarding nets, while telling them to be cautious as the enemy would be driven to desperation. And would you believe it, they left their blockade [at Nottawasaga] after only five days, separated from each other and lay carelessly each night at anchor?"

Chapter 17

JAMES FENIMORE COOPER
ENTERS THE DEBATE

Elliott's Last Stand

In conversation with officers of the Queen Charlotte *I asked why they had directed fire wholly upon* Lawrence *instead of* Niagara. *They replied because* Niagara *was so far off we could not injure her.*
—*Usher Parsons, letter of July 1818*

If Jesse Elliott had one distinct advantage in his dispute with Oliver Hazard Perry, it was that he outlived him by several decades. Unfortunately for Elliott, however, his ham-handed approach to command continued to land him in difficulty with the official naval administration. Perry's death seemed to energize his supporters, and in the years immediately following his demise while fighting pirates off Venezuela, even more towns, buildings, hospitals, schools and monuments were dedicated in his name. Ten American counties were named for Perry during the era just after his death. And every time something else was memorialized, the Perry-Elliott controversy would resurface in the national press.

By the mid-1830s, Elliott was in trouble again after a Mediterranean assignment where the navy had hoped to hide him. After a dispute with junior officers, there was a court-martial, and Elliott was suspended without pay and sent home to the United States. During his time in his hometown of Hagerstown, Maryland, friends urged him to write his own memoirs and tell his side of the Battle of Lake Erie. While not a gifted writer or orator, Elliott agreed to prepare a paper, which he finally delivered to his friends and supporters in Maryland. His motives in doing so may have been to try to salvage his reputation so that he could be reinstated. More than anything else, the fiery Scotsman was a loyal navy man, and he wanted to return to duty.

Elliott's work progressed at a painfully slow pace as his supporters were reading and editing. As they examined Elliott's side of the story, they began to lobby for their friend and hero. While the Hagerstown friends of Jesse Elliott were not as powerful as Perry's New England supporters, their efforts ultimately reached the attention of noted American writer James Fenimore Cooper, who was intrigued by the Perry-Elliott conflict. Cooper had already noted that Perry misused the famous words of Perry's friend and mentor James Lawrence when he crafted his famous battle flag that said "Don't Give Up the Ship." According to Cooper, the actual quote that Perry "adjusted" was "Never Strike the Flag of My Ship." Interesting and brave, perhaps, but according to Cooper, less patriotic. Cooper's revelation of this fact, which he gleaned by interviewing sailors who had been aboard the *Lawrence*, did not endear him to Perry supporters.

In 1839, while Elliott's manuscript was enduring more adjustments, James Fenimore Cooper examined Elliott's story, read the proceedings of the Barclay court-martial in England and interviewed several sailors who had been aboard the American ships during the battle. When he concluded his research, Cooper decided that it was Elliott who had been responsible for winning the Battle of Lake Erie and that his reputation had been tarnished by Perry and his supporters. Cooper's book caused a firestorm of controversy. He was branded a maverick historian, and for a time, his books were banished in the New York City Library System—not just the new, controversial book about Perry but all of his books. Perry supporters argued that James Fenimore Cooper had a bias against Perry that infected his objectivity.

Cooper was in a unique position to comment on naval history. Not only was he a prolific American author known for such sagas as *The Last of the Mohicans*, but he was also an academically trained (Yale) writer who had served in the United States Navy and seen sea duty. Critics, who seemed to be antagonized by his support of Elliott, noted that he had been expelled from Yale and that his writing was done in the style of a romantic novelist rather than a historian. Cooper responded that he was meticulous about the details of history, always interviewing participants and seeking firsthand testimony. Critics countered that Cooper's approach was indeed his biggest problem. He tended to become infatuated by the stories of his subjects and to celebrate them, sometimes beyond objectivity. The history of the United States Navy was in its infancy in those days, however, and Cooper supporters argued that the American writer was adding texture and humanity to important stories about historical events.

When James Fenimore Cooper was sued by outraged Perry supporters who claimed, among other things, that the maverick American-Scot was defending a fellow Scotsman (Elliott), the author answered their actions by countersuing. In each of four lawsuits, the author/historian was deemed innocent. Eventually, his classic books on naval history became standards at the United States Naval Academy.

The 1939 publication of Cooper's *History of the Navy*, which attempted to vindicate Elliott, motivated Hagerstown's hometown hero to finally present his own paper in Maryland in 1843. The paper was subsequently transcribed and published. In it, Elliott noted that he was originally guiding *Niagara* into battle in the first (in line) position, while Perry, aboard *Lawrence*, was in third position behind *Caledonia*. The battle plan, as outlined by Perry, was for the first three American ships to each engage one of the three large ships in the British battle line. This was traditional naval battle protocol at the time. Minutes before the first shot was fired by the British, Perry was asking megaphone questions about which British commanders were on the specific ships that were approaching. When he learned where Barclay was in the opposing British line, Perry insisted on switching positions with Elliott.

Given the light winds, Elliott said that his last-minute sailing maneuver that allowed Perry to pass put *Niagara* in irons behind *Caledonia*, where he was unable to maneuver or fire. Perry's passing maneuver, on the other hand, gave the *Lawrence* a burst of speed, which propelled it toward the British line in a dying wind. While Perry was moving toward harm's way and Elliott was stalled behind *Caledonia*, the first British shots were fired.

The British, according to Elliott, probably had their own similar battle plan, but since Perry alone sailed into range of the three large enemy ships, all of them commenced firing at the *Lawrence* at the same time. By the time Elliott could gain momentum by passing *Caledonia*, it was obvious that Perry and the *Lawrence* were doomed. Instead of following Perry directly into the mayhem, Elliott and the *Niagara* continued, with *Caledonia*, to batter the British with twelve-pound cannons. Elliott added that the *Lawrence* was unable to sail closer to the British ships because of the dying winds. When Elliott began to run low on twelve-pound cannonballs, he sent his longboat to fetch additional ammunition from the *Lawrence*.

Elliott noted in his own retrospective paper that the *Lawrence* gunnery crew, through no fault of Perry's, was lacking in skill and had done more internal damage to themselves and the *Lawrence* during the opening salvo of the battle than the British had managed. Elliott's crew had been together for more than a year and was far more skillful than Perry's. This was the original reason that Elliott had been placed first in line for the battle. When a longboat returned, it wasn't Elliott's boat with cannonballs. It was Perry's longboat from the *Lawrence*, and it held Perry himself, who was, according to Elliott, agitated and in a state of despair. According to Elliott, Perry said, "Cut all to pieces. The victory is lost. Everything is gone," as he was being helped aboard *Niagara*. To encourage his commander, Elliott responded that the battle was not lost and reportedly stated, "No sir, victory is yet on our side. I have a most judicious position and my shots are taking great effect. You take my battery and I will bring up the gunboats." According to Elliott, while the British had been focused on *Lawrence*, the *Niagara*'s and *Caledonia*'s better-trained gunnery crews had been systematically destroying the British ships.

It was Elliott, according to his own story, who asked Perry to take over the *Niagara* while Elliott climbed into the longboat and led the small American gunboats into the midst of the British fleet. Elliott went to great lengths in his paper to defend Perry as being a heroic and noble commander, even though he had previously been so angry with him that he had challenged him to a duel. Elliott added that after Perry's death, his vitriol toward his old commander had calmed. He also suggested that the battle sketches indicating the positions of the ships were inaccurate. In discussing Perry's accusations, Elliott referred to testimony following the battle, during which Perry had said under oath when asked if Elliott was to blame for not following him into battle, "No Sir, with her position when the battle commenced and the winds he had to contend with, no officer could have done better than Elliott did." When asked if it was indeed Perry's actions aboard *Niagara* that had won the battle, Perry responded, "The result would have been the same if I had not taken command of *Niagara*." Elliott noted that, in retrospect, it seemed as if it was Perry's younger brother who had caused much of the post-battle controversy instead of Perry himself.

In reviewing Elliott's actions, James Fenimore Cooper also used Usher Parsons' pro-Perry testimony to support his position in favor of Elliott. Parsons wrote in 1818, "After the action, I attended the wounded aboard *Niagara* and of twenty, only one or two told me that they had been wounded while Elliott was in command. Onboard the gunboats that Elliott later

Modern version of Oliver Hazard Perry's battle flag flying from the reconstructed brig *Niagara. Photograph by Jerry Skrypzak.*

brought up on the British, no more than two or three were wounded. The number killed or wounded aboard *Lawrence* was 83." While Usher Parsons, who spent much of the battle below deck, clearly interpreted this as reflecting bravery on the part of Perry, James Fenimore Cooper called that efficiency of manpower a superior battle tactic on the part of Elliott.

Pressure from Elliott's Hagerstown friends, in combination with Cooper's book, influenced President Tyler to reinstate him. Elliott left Hagerstown and took over the Philadelphia Naval Yard. In later years, Theodore Roosevelt wrote his own comprehensive book on the naval battles of the War of 1812, in which he reinforced Cooper's version of the Battle of Lake Erie but seemed to take a middle position with respect to the Perry-Elliott controversy. Given the volumes of material generated by various inquests and investigations, it would have been possible to assemble sets of quotes from firsthand observers that could have been used to support either the Perry or the Elliott position. Roosevelt's approach seemed to be an attempt to learn from past experience rather than to assign blame.

Chapter 18

SALVAGING PERRY'S FLEET

The Resurrections of Lawrence *and* Niagara

Left Black Rock after eight months. Commanders of the five vessels were Perry of the Caledonia, *Dobbins of the* Ohio, *Holdup of the* Trippe, *Almy of the* Somers, *and Darling of the* Amelia.
—*Usher Parsons' diary, June 12, 1813, as he first moved to Misery Bay*

Patriotic celebrations became an important component of United States life after the Civil War, and nothing was more likely to stimulate an event than an anniversary date. During 1875, in anticipation of the 100th birthday of the country, a group of Philadelphia preservationists began to make inquiries about the *Lawrence*. Locals surveyed the Misery Bay site of Perry's scuttled naval ships and declared the bones of the *Lawrence* to be essentially sound. By the late 1800s, Misery Bay had evolved into an industrial area that served Erie's downtown boat-building and commercial fishing interests. During those days of wooden ships, it was common practice to winterize ships by scuttling them after each summer shipping season. Sinking a wooden ship in shallow water slowed the process of rot and helped to preserve the wood.

By the 1870s, the waters of Misery Bay were literally jammed with scuttled ships. *Niagara* and *Lawrence* had been resting there for decades, and with the advent of steamships, they were joined by outdated schooners that had been judged too small to be converted to barges or steamers. In addition to the infusion of schooners, the 1871 closing of the canal between Erie and Pittsburgh led to a number of packet boats being scuttled in the same area.

The possibility of freeing up space by removing the *Lawrence* appealed to the locals who were running the Misery Bay operation, so as soon as they were commissioned, they went to work pulling the bones of Perry's flagship out of the mud. The *Lawrence* was relatively close to shore, and its retrieval went quickly. By the early summer of 1875, it had been loaded onto a railroad car and sent to Philadelphia. The *Lawrence* was not restored in the sense that the *Niagara* would ultimately be. Its hull was cleaned up and enhanced in a way that would allow observers to visualize its original size and shape. Most imagined that after the 1876 Centennial it would be sawed into pieces and offered as souvenirs.

The *Lawrence* was the hit of Philadelphia's 1876 Centennial. The remains of the ship were placed in an ornate pavilion at the fairgrounds and equipped with exhibits that explained the role of the ship in the War of 1812 and the importance of the war itself in helping to define America's place as a world power. Sadly, the exhibit caught fire in December, after the fair had closed. By the time anyone noticed the smoke, the *Lawrence* was gone. All that was left when firefighters reached the pavilion was ashes.

Several years later, some Erie people decided that it might be useful to repeat the experience of the *Lawrence* and raise the *Niagara* in time for the 100th anniversary of the Battle of Lake Erie. By that time, *Niagara*

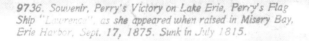

The hull of the *Lawrence*, Perry's original flagship, raised from Misery Bay in 1875. *Jerry Skrypzak Collection.*

Steam barge lifting the hull of the *Niagara* through the frozen surface of Misery Bay in 1913. *Jerry Skrypzak Collection.*

was owned by the Dobbins family, who were happy to sell it to the local preservation committee that had decided to pursue the project. Erie shipbuilder William Paasch was commissioned to direct the project, and he, in turn, hired local boat builder Herman Lund and his uncle Hans Hansen to assist. The restoration of *Niagara* was based on plans that had been produced by Howard Chapelle, who based his plans on the general work of Noah Brown.

Niagara was resting in deeper water, however, so the task of raising it was more challenging than dragging out the *Lawrence* had been thirty years earlier. After surveying the underwater wreck, Paasch attached floating winter poles to four points on the hull during the fall of 1912. Then, in March 1913, he cut a hole in the ice following the contour defined by the winter poles. Using a steam barge, he winched the hull up though the ice, and when *Niagara*'s hulk had been raised through the surface, winches were used to slide it to shore.

Niagara was repaired and re-rigged during the spring and summer, but the work took longer than had been anticipated. In mid-May, seeing that the anticipated launch date was at risk, the local committee asked Erie's Hammermill Paper Company for help. Always community-minded, the company provided three riggers who worked seven days a week with the Paasch crew until the brig was launched on June 7, 1913.

Niagara's hulk after being winched to shore across the frozen surface of Misery Bay in 1913. *Jerry Skrypzak Collection.*

Niagara's rigging crew shown at Misery Bay in 1913. *Jerry Skrypzak Collection.*

To celebrate the 100th birthday of the Battle of Lake Erie, *Niagara* went on a Great Lakes tour, becoming an ambassador for the city of Erie. The rebuilt ship returned to gala local celebrations in late September and was berthed at the Public Dock. Within a few years, however, competition for space by commercial fishermen resulted in the ship being sent back to Misery Bay. After a few winters of neglect, signs of the rushed rebuild began to show. *Niagara* was leaking badly and in danger of sinking into the mud again. In 1915, the ship was cribbed up near shore and lifted out of the water.

An interesting local controversy emerged when the decision was originally made to remove the brig from Misery Bay. A vigorous local cottage industry had evolved during the early 1900s, in which woodworkers stripped materials from the sunken ship and used it to carve souvenir items. Letter openers, gavels, desk accessories, walking sticks and other trinkets were regularly carved from relic wood of the *Niagara* and sold for premium prices. In consideration of the number of such items that still circulate among collectors, as compared to the size of the ship itself, some have wondered if all of the items that were said to have come from the ship are authentic *Niagara* relics.

The restored brig *Niagara* about to be launched in 1913. *Jerry Skrypzak Collection.*

Above: *Niagara* on her triumphant 1913 Great Lakes tour being towed into Toledo, Ohio, by the tug *Buffalo*. *Jerry Skrypzak Collection*.

Left: Crowds visit the restored *Niagara* at Erie in 1913. *Jerry Skrypzak Collection*.

The neglected *Niagara* deteriorating at the foot of State Street in the 1940s. *Jerry Skrypzak Collection.*

The brig *Niagara* sat out of the water deteriorating until 1929, when a new restoration began but quickly stalled for lack of funds. Herman Lund, who had done much of the 1913 restoration, was placed in charge in 1933 and moved the ship to today's *Niagara* launch area on Presque Isle, where he began again. Lund's choice of the new work site was driven by the fact that Presque Isle had become a state park in 1928 and officials were happy to make a place for the historic brig and its restoration. Lund's funds ran out after a year, but President Franklin D. Roosevelt, a ship and model lover, came to the rescue with $30,000 in WPA funds. The restoration was completed in 1943, and *Niagara* was launched from Presque Isle and moved back to the Public Dock.

By 1945, *Niagara* was sinking again at its downtown berth. To rescue the ship, it was lifted out of the water and placed in a concrete exhibition cradle on Erie's lower State Street. To celebrate the sesquicentennial of the Battle of Lake Erie, *Niagara* was refurbished in its cradle and re-rigged in 1963, where it continued to sit, welcoming visitors for the next 45 years.

Chapter 19

WASAGA BEACH GIVES UP ITS DEAD

The Nancy Rises from the Silt

Bold rocky coast, boisterous, covered with impenetrable fogs, filled with small islands, and sunken rocks.
—Arthur Sinclair's description of the North Country

During the 1900s, Ontario's Georgian Bay region evolved into one of North America's premier tourist areas. The growth of Hamilton and Toronto, combined with American tourism from Michigan and other northern states, brought thousands of cottagers and tourists into the area. The rugged beauty of the islands and waters between Tobermory and Lions Head on the western shore and the thousands of islands on the eastern shore provided stunning vistas of the north. The stark beauty of the region continues into Georgian Bay's North Country, where the LaCloche Mountains frame the North Channel and Killarney Provincial Park.

At the southern extremity of the Georgian Bay, near Miller Worsley's old Nottawasaga River Basin, a thriving Provincial Park now marks the broad beach that once separated the open waters of the bay from the river. A modern tourist area called Wasaga Beach has evolved there, attracting restaurants, shops, cottages and homes in addition to the provincial park.

Explorations of the Nottawasaga River led to an amazing discovery in 1911. There, silted into the bottom of the river, was the hull of an early 1800s-era sailing ship that local preservationists immediately suspected to be the *Nancy*. Not much had been said about the old ship since its

Aerial view of today's Wasaga Beach and the Nottawasaga River showing the Nancy Museum. *David Brunelle, Laurel Finney and Wasaga Beach Provincial Park.*

disappearance years before people had been attracted to this area of Ontario. Most had assumed that the fire followed by the ravages of time had destroyed any vestiges of the ship. But given the rich history of the region and the continuous maritime activity at Penetanguishene, Ontario, just a few miles north, there was interest in raising and preserving the ship. If the War of 1812 had not ended, Penetanguishene would have become the site where the British navy was to build its fleet for an 1815 Campaign, and in 1817, the British established a naval base there.

Summer seasons are short on the Georgian Bay, and lacking the resources to continue with a proper preservation plan, locals soon forgot about the set of bones that seemed so likely to have been the *Nancy*. But in 1924, an amateur archaeologist discovered a cannonball buried in the sand near the underwater remains of the *Nancy*. After study, it was determined that the projectile was an American thirty-two-pound cannonball and that it almost certainly had to have been fired from a *Niagara* carronade during the 1814 attack at Nottawasaga Bay.

The discovery of the cannonball led to a resurgence in interest in the previously discovered, but largely ignored, bones of the *Nancy*. In 1928, preservationists using grants from both the Canadian federal and Ontario

Contemporary photograph of Discovery Harbour at Penetanguishene. *Kathleen Trainor and Huronia Historical Parks.*

The Nancy Museum at Wasaga beach. *David Brunelle and Wasaga Beach Provincial Park.*

Preserved hull of the *Nancy* at the Nancy Museum in Wasaga Beach, Ontario. *David Brunelle, Laurel Finney and Wasaga Beach Provincial Park.*

Top: Recovered cannonball from the 1814 attack on the *Nancy* by *Niagara*. Photograph by Jerry Skrypzak and John Baker.

Bottom: Figurehead from the *Nancy*. David Brunelle, Laurel Finney and Wasaga Beach Provincial Park.

provincial governments raised the hull of the *Nancy* and moved it into a museum that was especially built to preserve and display this important piece of Canadian and American history. The preservationists who worked tirelessly to make the *Nancy* project happen were partially inspired by the work that had been accomplished in Erie just two decades earlier in raising the *Niagara* and taking it on the famous 1913 Great Lakes tour.

The Nancy Museum, which opened its doors to visitors a few months after the hull was raised, continues to attract history buffs who are excited about tracing the events of the War of 1812. The schooner *Nancy* historical site features a walking bridge that allows visitors to stroll a few steps to Nancy Island, a land form that was created by the silt and sand that was diverted from the underwater wreckage. The Nancy Island site is now home to one of the largest annual War of 1812 reenactment events in North America.

During the early 2000s, the actual cannonball that had been responsible for much of the development of the Nancy Museum was presented to Walter Rybka, captain of the *Niagara*, while the brig was on tour in the Georgian Bay. The cannonball is currently on display at the Erie Maritime Museum.

Chapter 20

THE NEW *NIAGARA*

The Brig Sails Again

To be effective as a proper piece of living history, the Niagara *must sail.*
—*Captain Walter Rybka*

By the 1980s, the venerable old *Niagara* was beginning to look shabby once more. Sitting out of the water in its dingy concrete cradle on lower State Street, it seemed a sub-optimal symbol of Erie, the "Flagship City." Finally, a group of locals decided that something had to be done. After a series of meetings and discussions, the Niagara League was formed and quickly developed an ambitious plan to rebuild the old ship and then focus on making it the centerpiece of community renewal. The new *Niagara* would highlight Erie's heritage as the home port of Perry's Battle of Lake Erie fleet. The plan was to restore *Niagara* in a way that would allow it to sail again, not just to float at its berth for a few years or be towed on a multi-city celebratory tour, as it had in 1913. This bold new vision involved the creation of a piece of living, sailing history.

In 1986, after fundraising and a grant from the Pennsylvania Historic and Museum Commission (PHMC), Melbourne Smith was hired to lead the project. Smith returned to the 1913 plans that had been developed by Howard Chapelle and based on the work of Noah Brown, *Niagara*'s original builder. Smith hired teams of experienced traditional boat builders who began their work in 1987. The old brig was moved to a secure location and disassembled, but the damage and rot was so severe that little of the earlier restoration could be salvaged.

Niagara languishes on Erie's Lower State Street in her concrete cradle about 1980. *Photograph by Jerry Skrypzak.*

A literal beehive of activity erupted on the bay front, with traditional wooden boat craftsmen setting up thirty foot-steamers and building a new keel and rib system following Chapelle's plans. This time the construction utilized properly seasoned yellow pine and spruce, and no corners were cut. With excited spectators watching and wondering, twenty-foot boards were steamed until they became malleable, and then they were lifted into place on *Niagara*'s hull. Working in teams, the builders skillfully shaped the steamed boards into *Niagara*'s compound curves while spectators commented that the new brig *Niagara* would almost certainly be an improvement over the original or any of its restored cousins. Even with all the new technology that was available, the rebuild took significantly longer than the construction of the original 1813 *Niagara* or the 1913 restoration.

The newest *Niagara* was launched in 1988, in time to celebrate the 175th anniversary of the Battle of Lake Erie. After two years of rigging, sea trials were conducted in July 1990. The ship was engineless and had to be towed out into the lake for its first test. An armada of sailing ships, including many of the city's fastest club sailboat racers, followed to see what she could do. Skeptics joked that the ship was too high, too beamy and not well rigged. Few expected it to perform at a high level. Aboard *Niagara*, however, the mostly volunteer crew, making its first exploratory test of the sails and rig,

Builders begin to restore the *Niagara* to sailing condition in 1987. *Photograph by Jerry Skrypzak.*

Restored *Niagara* being launched with a crane in 1988. *Photograph by Jerry Skrypzak.*

raised an almost full brace of canvas and left the flabbergasted spectator boats in its wake. *Niagara* was a rousing success.

Two additional bits of municipal infrastructure were required for the new brig to be the success that it has become: a proper home and a roadway system capable of bringing visitors to see and experience it. After much community discussion, a decommissioned, coal-fired electric power plant on the waterfront was converted to a beautiful and modern combination public library and maritime museum. The museum portion of the development was designed to house the *Niagara* and its maintenance operations. The resulting Erie Maritime Museum opened in 1998. The next step was the development of a highway infrastructure that linked the waterfront home of *Niagara* with major throughway systems, allowing visitors easy access to the site.

An experienced tall-ship captain, Walter Rybka, was recruited to take charge of *Niagara* and its operations in 1991, and within a few years, he had developed sailing programs, a combination paid and volunteer crew and a sailing school. Most importantly, he initiated a schedule of summer sailing trips that returned *Niagara* to its former role as ambassador for the city of Erie and the state of Pennsylvania. Modern diesel engines were ultimately

The Erie Maritime Museum. *Photograph by Jerry Skrypzak.*

Above: The restored
Niagara under sail.
*Photograph by Jerry
Skrypzak.*

Left: Captain Walter
Rybka of the *Niagara*.

The restored brig *Niagara* at her berth near the Erie Maritime Museum. *Photograph by Jerry Skrypzak.*

added to make the ship self-sufficient as it traveled, and programs of local day sails were offered to give visitors the opportunity to experience life aboard a tall ship.

After several years, I finally had the opportunity to go sailing aboard the brig. By this time, I had been a longtime racing and cruising sailor and had logged thousands of offshore hours sailing on the Great Lakes as well as the oceans. While the sailing itself didn't surprise me, the experience of being aboard *Niagara* when the sails were raised and the ship dipped her rails toward the sea and accelerated through the water gave me goose bumps. It was a living thing whose sounds and motions instantly reconnected me with everything that I had learned and read about the grand age of sail. *Niagara* was a history experience to be felt rather than read or discussed—so much more than a label on a museum exhibit.

In the years since its re-launching, *Niagara* has become a member of the Erie community. Its presence is a vivid reminder of the birth of the city and an invitation for locals to become involved. A volunteer support group called the Niagara League has helped to present and interpret the brig.

Chapter 21

SHIPWRECK IN LAKE ERIE

Could It Be Caledonia?

Scorpion *returned from lake Huron to tell us that the enemy planned to meet us* *on the lake in force. On account of that we recalled* Caledonia *which had been dispatched back to Erie.* —Usher Parsons' diary, *July 22, 1814, on the way to Mackinac*

For several years, rumors of a War of 1812 shipwreck lying in mid–Lake Erie, east of the city of Erie, have been circulating among divers. Perry's fleet included two ships (including the *Ohio*) that had been moved to Erie from Black Rock, New York, after being captured by Jesse Elliott. One of those ships was the *Caledonia*, a 180-ton brig that had originally been a British merchant ship owned by the British North West Company. *Caledonia* was built in 1807 at Amherstburg, where it was designed for the Georgian Bay fur trade. The *Caledonia* was commandeered for the war in 1812 and equipped with two twenty-four-pound long guns and a single thirty-two-pound carronade. In October 1812, the HMS *Caledonia* was captured by Americans near Fort Erie, moved to Black Rock (Buffalo) and renamed USS *Caledonia.*

The *Caledonia* played an important role in the Battle of Lake Erie. Making the best use of its long guns, its captain positioned the ship almost a half mile from the main skirmish and continued to pound away at the British ships as the battle ensued. When the battle was over and Perry returned to Erie to set up his Misery Bay naval base, many of the ships, in particular the *Lawrence*, were so badly damaged that observers wondered why he had bothered to bring them back. Perry was sent to the Atlantic, but his bedraggled Misery Bay fleet continued to deteriorate in Erie. The USS *Caledonia*, which had been one of the least damaged ships, participated in the 1814 Campaign and returned to Misery Bay once more for the winter of 1814–15.

By the winter of 1815, however, it was apparent that the Erie Naval Base and its fleet had outlived their usefulness. With the war settled, the United States Navy ultimately closed its Erie base and sold the Misery Bay yard and its contents to businessmen who hoped to salvage some of the ships and rebuild them for commercial use. The built-in-Erie military vessels were so hurriedly constructed that they were of little or no commercial value. In the rush to launch the fleet, Dobbins had used green (non-seasoned) wood, inferior softwood species (sassafras and elm) and wooden pins in place of nails. As a result, the bulk of the Misery Bay fleet was in deplorable condition.

Caledonia was the most likely candidate for resale as a commercial vessel, and in 1820, it was refitted and returned to its original configuration as a merchant vessel. Erie's shipping business was booming, and the refurbished *Caledonia* provided an expedient way for local merchants Rufus Reed and John Dickson to return to the salt and fur trade after the war. *Caledonia* was renamed *General Wayne* and placed into service. Reed passed his share of the ship to his son Charles in the 1820s. Charles Reed later bought his partner's share when Dickson moved from Erie to Meadville in 1929. By the mid-1830s, Charles Reed had reconfigured his shipping business toward steamships and abandoned the acquisition or repair of sailing ships. His thirty-two-ship fleet, as of 1840, included seven sailing vessels, but the *Caledonia/General Wayne* was not listed among them.

When Reed's sailing schooners reached the end of their productive days, he ordinarily dismantled them and salvaged the parts. Since Reed's need for ships far exceeded the rate at which he could build or acquire them, he sometimes installed steam engines and boilers in the biggest and best of his schooners to try to extend their lives. But he did not convert the venerable old *General Wayne* (*Caledonia*), which by that time was almost forty years old. When Reed launched his flagship steamer *Erie* in 1837, he announced that he was "through with sailing vessels."

Official shipping records suggest that Reed's *General Wayne/Caledonia* worked into the early 1830s and then disappeared. There are only three plausible explanations for its disappearance: (1) It may have sunk by accident, although such an event would have received news coverage; (2) It could have been sunk or scuttled on propose; or (3) It may have been dismantled and salvaged. Since the news of the day was filled with reports of vessel traffic and a sinking was always headline material, it seems unlikely that *Caledonia* was lost at sea. As to the likelihood that it was scuttled offshore, that would have been unlike the thrifty Reeds, who meticulously dismantled their unneeded ships and stored or sold the salvageable parts. It was common for Charles Reed to tie a boat to his pier and consider its fate for a year or two when it had reached the end of its life. That might explain the apparent uncertainty regarding the final year(s) of its use.

In 2004, a Massachusetts salvage group called Northeast Research announced the discovery of a two-masted schooner lying in 175 feet of water at the bottom of Lake Erie between Dunkirk, New York, which is east of Erie, and Long Point on the Ontario shore. Underwater photographs of the ship bear the look of a typical pre–Civil War, working schooner, and its hold contained grain and hickory nuts, both common schooner cargos during the early to mid-1800s. After several dives, the salvage group, which was operating out of Dunkirk, made an official claim to the wreck and produced artifacts that had been recovered. In its claim, the dive group argued that the wreck in question was in fact *Caledonia* and that it was of historical significance because it had been part of the American fleet during the Battle of Lake Erie.

Northeast Research pitched a clever vision of the salvaged *Caledonia* exhibited in a huge aquarium and preserved by immersion in cold water. Instead of approaching the relatively small town of Dunkirk, which is the closest American city to the wreck but would probably not have had the financial resources to mount such an endeavor, the group took its idea to Buffalo.

When they argued their case for ownership before a state magistrate, the members of the dive group were chastised for removing artifacts and for the possible mishandling of human remains. Even though Northeast Research produced a letter from one of the "owners" of the *General Wayne* (a surviving member of the Reed family) stating that they had permission to salvage the ship, the court concluded that the shipwreck was the property of the State of New York and should not be disturbed. The court also noted that the salvers had not really established that the wreck in question was, indeed, the *Caledonia/General Wayne*, and therefore it had no particular historical significance. Several "experts" for the state testified that the shipwreck seemed more likely to be a 1840s-era vessel than an early 1800s design.

The magisterial decision was appealed to a federal court. In the appeal, the salvers argued that even if the ship was not the original *Caledonia*, it

Suspected wreck of the *Caledonia* on the bottom of Lake Erie east of Erie. *Courtesy of* X-Ray Magazine.

was likely to have been involved in transporting slaves to safety during the underground railroad era preceding the Civil War. Thus, even if it was an 1840s-era design, it might still be of great historical significance. In August 2010, Judge Leslie Foschio upheld the earlier state ruling that the ship belonged to New York State and that it should remain where it was, on the bottom of Lake Erie. The judge ruled that the salvers had inappropriately removed artifacts, damaged the ship during early dives and mishandled possible human remains that were reported to have been aboard. Judge Foschio also noted that the objective of raising and preserving the ship was misguided and unrealistic and that any such efforts would be more likely to do harm than good to the wreck.

In sorting through the apparent facts regarding the unknown shipwreck in Lake Erie, one more bit of evidence should be noted. In 1934, Canadian historian George Cuthbertson published an article about *Caledonia* in the Canadian historical journal *Beaver*. In his research, Cuthbertson used primary data obtained from the business records of Charles Reed's Erie shipping company to chronicle the end days of the *Caledonia/General Wayne*. Cuthbertson noted that the Reeds followed their traditional business practices and dismantled the old ship after it had outlived its usefulness, become too expensive to repair and proved structurally unable to tolerate the addition of a steam engine. According to Reed Company records, salvaged parts of *Caledonia*'s hull and rigging were sold for firewood and scrap iron.

While it seems likely that the wreck in question is not *Caledonia*, no one will ever know for sure. Perhaps the notion of raising a fragile mid-nineteenth-century shipwreck from the bottom of Lake Erie and preserving it in an aquarium for Buffalo tourists to visit was overly ambitious and misguided, but Northeast Research's project was not entirely without merit. The timing of the *Caledonia* story was especially interesting in light of the upcoming two-hundred-year anniversary of the Battle of Lake Erie, and the story of the nonmilitary brig would not have been so well told without their efforts. And for Lake Erie divers, who generally refer to it as the "Admiralty Wreck," speculation will continue as summer expeditions take them to inspect the mysterious two-masted shipwreck.

Chapter 22

THE MYSTERIES
OF HISTORY

Approaching the Two-Hundred-Year Anniversary

*At 10 O'clock (just before the battle) martial music stuck up the thrilling sounds
of All Hands to Quarters. The fighting flag was then displayed at mast head and
the valor and patriotism of the crew appealed to by the commodore, which they
responded to with three hearty cheers.*
—*Usher Parsons' letter to the son of Oliver Perry*

I grew up in Erie, Pennsylvania, in the 1950s. Like countless grade-
schoolers my age, one of my favorite class field trips was our annual
visit to the concrete-encapsulated brig *Niagara* downtown. We kids didn't
care or even appreciate that the ship was not in the water. It was a ship
and a day off school, and it was amazing. Once aboard, we bent over
and scurried through the below-deck compartments, marveling at "how
short the old sailors must have been." Exuberant docents scared us in the
darkened quarters with loud grunts and screams and yelled "Arrr" and
"Avast, mateys" at the tops of their lungs. On *Niagara*'s deck, we were
regaled with frightening stories of bloody body bits being slammed against
masts and other stylized tales about the bravery of Commodore Oliver
Hazard Perry.

In our hometown, we had a Perry School and a Perry Township with
its Perry Fire Company, not to mention our very own Perry Monument, a
slightly shorter replica of the Perry Monument at Put-in-Bay. We also had

The last remaining Erie artifact from Perry's Westside *Lawrence* and *Niagara* shipyard is this stone memorial. *Photograph by Jerry Skrypzak.*

dozens of businesses beginning with the word Perry, as well as a downtown park named Perry Square. We clearly understood the importance of the name. Most of us probably failed to comprehend some of the important historic nuances. Some kids were left with the impression that the Battle of Lake Erie had taken place right here in Erie, and why not—that was where the *Niagara* was. Others were convinced that *Niagara* had been carrying escaped slaves to Canada as part of the Underground Railroad or that the brig fought in the Civil War. Regardless of our misunderstandings, however, we were all left with one important reality: Perry was the hero who put Erie, Pennsylvania, our town, on the map.

While writing previous books about Great Lakes ships, shipping and shipwrecks, I have avoided the general topic of the Battle of Lake Erie. I knew something of the post-battle controversy between Perry and Elliott and didn't want to have to take a side. Some years ago, I spent a weekend in a Buffalo library reading the collected papers of Daniel Dobbins and was somewhat surprised to learn the depth of the old shipwright's disdain for Perry. Realizing the fiery nature of Dobbins, however, I discounted much of what I was reading and focused primarily

Entrance to Dobbins Landing at Erie showing the commemorative placard. *Photograph by Jerry Skrypzak.*

Aerial view of Dobbins Landing, Erie's public dock. *Photograph by Jerry Skrypzak.*

on Dobbins' accomplishments as a sailing master. After all, the town dock is named for him.

I also recall my college days at Erie's Gannon University when locals were celebrating the sesquicentennial of the Battle of Lake Erie. My most vivid recollection of that summer event, other than taking my girlfriend (who is now my wife) to the parade, was that many of Erie's men celebrated by growing beards. Not sure of the significance of the facial hair–growing, I gave it a "college try," only to be criticized by my professors for such foolishness. It was 1963, several years before longhaired hippies took over the nation's colleges and universities.

As the year 2013 approached, I found myself being asked to prepare something about the Battle of Lake Erie. During the summer of 2013, our city will literally explode with 200th anniversary celebrations. My work began

Sesquicentennial logo from Erie's 1963 year of celebration. *Bob Zawadzki Collection.*

Perry Monument at Erie's Misery Bay. *Photograph by John Baker.*

with a simple power point presentation, which I imagined to be the extent of my contributions. To test the power point ideas, I decided to ask the Jefferson Educational Society of Erie if it would be interested in having me speak to its members. Erie's Jefferson Society is a think tank dedicated to community growth, development and celebration, and it is directed by one of the most talented historians that I know, Dr. William Garvey. The society is the primary sponsor for the upcoming year of celebration, which is to include a tall-ship parade in the local harbor.

In thinking about the power point presentation, which I imagined that Dr. Garvey would be attending, I felt more than a bit of pressure to enrich the slides with research. While I am not a historian by training, I am an academic researcher who has had the benefit of working with talented historians who took time to help me understand qualitative research methods. My own training, which has been in organizational psychology and leadership, is not entirely disconnected from history. In fact, I try to imagine myself as what my favorite science fiction author, Isaac Asimov, called a psychohistorian, a person with interdisciplinary training who tries to use more than one frame of reference to understand history. Undaunted, I plunged into the primary data that described the events and characters that were shaped by the Battle of Lake Erie. As I read, I decided to focus my research on the events that followed the actual battle, rather than the battle itself.

My investigations led me to the diary of Usher Parsons and his archived letters, the proceedings of the court-martial of Barclay, the papers of Jesse Elliott and the interpretations of James Fenimore Cooper. The psychologist in me immediately cast a skeptical eye toward all these characters. Parsons experienced the battle from below deck and could only have understood it from the perspective of Perry, whom he was trying to impress so that he could advance his naval career. The $225,000 prize money, which was enormously significant if understood in modern context, clearly influenced Dobbins, who was a salty character in his own right. Elliott was a mercurial Scotsman whose entire career was marked by his continued anti-bureaucratic outbursts. Perry seemed to be a blueblood Rhode Islander who was positioning himself, as some have speculated, for a possible political office. Reading Perry's written flip-flops in the years after the battle reminded me of more than a few modern politicians. Finally, there was James Fenimore Cooper, the non-historian/naval historian who was thrown out of Yale for such pranks as training a donkey to sit on his mentor-professor's chair. Given the various opinions of these characters, and the obvious biases that almost certainly drove their narratives, I, like other students of the battle, am left with confusion over who was the actual hero of the battle. Perhaps a better approach might be to credit both of them for what they did and forget about identifying "the" hero from the Battle of Lake Erie.

Perry lived and worked in Erie, and his fleet, which was built here, clearly shaped Erie forever. His action of sailing into the teeth of the British fleet to engage it with a raw, untrained crew at cannonade (short) range was clearly heroic. Elliott's failure to follow Perry into the close-quarter fight may have been clever strategy, sensible reluctance, cowardice or a combination of the three. Some naval historians argue that if Elliott had indeed followed Perry into the battle, the two American ships together would have been too much for the British, and American casualties would have been minimized. Others speculate that the last-minute shift in battle line positions between Perry and Elliott was a primary cause of the damage that the *Lawrence* experienced.

Elliott's later argument that he was stuck behind *Caledonia* and unable to get free to follow *Niagara* into the line of fire may or may not have been entirely accurate. That story, told by Elliott decades after the battle, would certainly have been his best form of cognitive dissonance. It is difficult to reconstruct that aspect of the battle in hindsight, particularly in light of the dying wind that seems to have left *Niagara* unable to sail efficiently toward the *Lawrence* and the British fleet. It has been noted in pro-Perry arguments that Elliott could have used the *Niagara*'s sweeps to follow the *Lawrence* into battle. Elliott raised another aspect relating to Perry's difficulties with the *Lawrence* and its gunnery crews. He noted Perry's willingness to allow Elliott to keep his experienced gunnery crew together on the *Niagara*. Most fleet commanders would have mixed the experienced crew from Elliott's ship with the inexperienced crew aboard *Lawrence*. This concession by Perry may have been part of his noted New England gentleman's generosity, a characteristic that he repeatedly displayed after the Battle of Lake Erie as he ministered to and celebrated the defeated enemy, including Barclay.

Walter Rybka, captain of the reconstructed *Niagara*, who has spent a career reading and thinking about these matters, notes that if Perry had been killed either in the battle or on the longboat ride to *Niagara* from the *Lawrence*, Elliott would almost certainly have become the acclaimed hero of the battle, and all the things that are now named after Perry would likely be called "Elliott" instead: Elliottville, Elliott County, the Elliott Monument, etc.

In my retrospective view, I continue to think that the best stories regarding Perry's fleet began after the famous battle in September 1813 and that the fleet may have been more important than any single individual, including Perry or Elliott. The stories of Usher Parsons, Daniel Dobbins, Miller Worsley and Arthur Sinclair add to and enrich the narrative that defines those tumultuous times, which some historians have described as America's Second War of Independence.

The year 2013 will mark the 200th anniversary of the Battle of Lake Erie, an event that continues to symbolize the meaning and life of the city of Erie. Like a dozen other Great Lakes cities and towns, Erie will be launching a year of celebrations, which will include concerts, parades, a tall-ship festival and other important activities. Erie's Jefferson Educational Society has initiated a two-year, multimillion-dollar program to guide and fund the celebration. One important initiative that is planned for the year 2013 is a renewed program of bringing local and regional history back into the schools. It seems that in the rush to cut costs and improve standardized test scores, the most glaring topical victim has been attention to local history. It seems that without the same "regular doses" of local folklore and heroes that kids my age were exposed to, today's youngsters may have lost touch with the roots that would make them feel pride in their hometown and its connection to the evolving history of North America. Sadly, the generation of young

Statue of Oliver Hazard Perry at Erie's Perry Square. *Photograph by Jerry Skrypzak.*

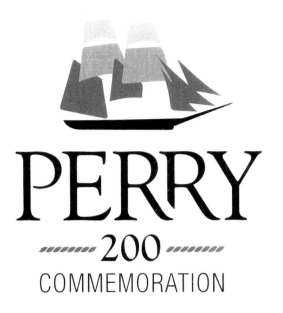

Left: The Jefferson Educational Society's "Perry 200 Commemoration" logo in anticipation of the year 2013 celebrations. *Dr. William Garvey and the Jefferson Educational Society*.

Below: The kids who currently visit and sail *Niagara* are the best hope for preserving the brig for the next one hundred years. *Photograph by Jerry Skrypzak*.

teachers that could otherwise be counted on to bring this important regional history to their classrooms is equally deficient in local history, which makes them uncomfortable in presenting this material.

It is my hope that this volume will add to the rich stories that will be told over the next years in Erie and other United States and Canadian cities that have a stake in the naval wars that occurred on the Great Lakes. And that somewhere, the influence of these stories will reach a new generation of children who will find themselves excited by the adventures of the heroes who manned Perry's fleet, from Perry himself to Jesse Elliott, Miller Worsley, Daniel Dobbins and Usher Parsons. In case I won't always be there in person to help tell all these stories or to act as a War of 1812 docent: "Arrr" and "Avast, mateys!"

Appendix

AMERICAN AND BRITISH SHIPS FROM THE BATTLE OF LAKE ERIE

AMERICAN				
Vessel	Rig	Rating (Tons)	Length (Feet)	Cannons/ Carronades
Ariel	Schooner	70	70	4/0
Caledonia	Brig	85	75	2/1
Lawrence	Brig	277	116	2/18
Niagara	Brig	277	116	2/18
Ohio	Schooner	62	80	2/0
Porcupine	Schooner	50	60	1/0
Scorpion	Schooner	60	60	1/1
Somers	Schooner	65	60	1/1
Tigress	Schooner	50	60	1/0
Trippe	Sloop	50	55	1/0

BRITISH				
Vessel	**Rig**	**Rating (Tons)**	**Length (Feet)**	**Cannons/ Carronades**
Chippewa	Schooner	35	65	1/0
Detroit	Corvette	300	112	17/2
Gen. Hunter	Brig	75	65	8/2
Lady Prevost	Brig	96	72	3/10
Little Belt	Sloop	60	60	1/2
Queen Charlotte	Corvette	200	99	3/14

The cannon category includes long guns, which were rated according to the size of the cannonballs fired from them and ranged from twelve to thirty-two pounds for the American fleet and from two to twenty-four pounds for the British. Carronades were designed for close fire. The American fleet held the advantage in carronades and the British in long guns.

BIBLIOGRAPHY

Altoff, Gerald. *Oliver Hazard Perry and the Battle of Lake Erie*. Put-in-Bay, OH: Perry Group, 1999.

Berton, Pierre. *Flames Across the Border*. Toronto: McClelland and Stewart, 1981.

———. *The Invasion of Canada*. Toronto: McClelland and Stewart, 1980.

Blanche, Burt. *Captain Robert H. Barclay*. Vol. 14. Toronto: Ontario Historical Society Records and Papers, 1916.

Chamberlain, Ryan. *Pistols, Politics and the Press*. New York: McFarland, 2008.

Cocke, John H. *Papers: Arthur Sinclair to John Cocke*. Alexandria: University of Virginia, Alderman Library, 1843.

Cooper, James Fenimore. *The History of the Navy of the United States of America*. Philadelphia: Lea Blanchard, 1839.

———. *Lives of Distinguished American Naval Officers*. N.p., 1856.

Dillon, Richard. *We Have Met the Enemy: Oliver H. Perry, Wilderness Commander*. New York: McGraw Hill, 1978.

Dobbins, Daniel. *Papers, War of 1812 Collection*. Buffalo, NY: Buffalo and Erie County Historical Society, 1854.

Dudley, William. *The Naval War of 1812 in Four Volumes*. Washington, D.C.: Naval History Center GPO, 1985.

Dutton, Charles. *Admiral Oliver Perry*. New York: Longhouse Green and Company, 1935.

Elliott, Jesse D. "The Address of Com. J.D. Elliott USN Delivered in Washington County Maryland on November 24, 1843." Philadelphia: G.B. Zieberd Co., 1844.

Fredricksen, John C., ed. *Surgeon of the Lakes*. Erie, PA: Erie County Historical Society, 2000.

Hickey, Donald R. *Don't Give up the Ship: Myths of the War of 1812*. Urbana: University of Illinois Press, 2006.

Ilsevich, Robert D. *Daniel Dobbins: Frontier Mariner*. Erie, PA: Erie County Historical Society, 1993.

Malcolmson, Robert, and Thomas Malcomson. *HMS* Detroit*: The Battle for Lake Erie*. Annapolis, MD: Naval Institute Press, 1991.

Parsons, Charles W. *Memoir of Usher Parsons M.D.* Cornell, NY: Cornell University Library Collection, 1870.

Paullin, Charles. *The Battle of Lake Erie*. Cleveland OH: Raufin Club, 1918.

Perry's Victory Centennial Souvenir Booklet. New York: American Journal of History, 1913.

Rhodes, James A. and Dean Jauchius *The Court Martial of Commodore Perry*. New York: Bobbs-Merrill, 1961

Roosevelt, Theodore. *The Naval War of 1812*. New York: Putnam, 1882.

Ryerse, Amelia Harris. In *The Loyalists of America and Their Ties, 1620 to 1816*, by Edgerton Ryerson. Toronto: Elden House Company, 1880.

Skaggs, David C., and Gerald T. Altoff. *Signal Victory*. Annapolis, MD: Naval Institute Press, 1997.

Stanley, George. *The War of 1812: Land Operations*. Toronto: Macmillan of Canada, 1983.

Welsh, William H., and David C. Skaggs. *War on the Great Lakes*. Kent, OH: Kent State University Press, 1991.

White, Patrick C.T. *A Nation on Trial*. New York: Wiley, 1965

INDEX

North West Trading Company 26, 63
Nottawasaga Bay 58, 60, 63, 67, 69, 70, 105
Nottawasaga River 58, 60, 104

O

Ohio 49, 52, 76, 77, 90, 116
Ojibwa Indians 25, 28, 43, 57, 58, 60, 69, 71
Onondaga (Iroquois) Indians 25
Ontario 87

P

Paasch, William 99
Parsons, Usher 16, 19, 20, 21, 23, 24, 29, 30, 32, 33, 37, 39, 40, 41, 50, 52, 60, 61, 66, 88, 89, 95, 96, 126, 127, 130
Penetanguishene, ON 43, 58, 63, 105
Perry 200th Commemoration 12, 124, 127
Perry Monument, Erie 121
Perry Monument, Put-in-Bay 121
Perry, Oliver H. 23, 29, 30, 33, 37, 48, 49, 50, 52, 75, 76, 80, 82, 83, 84, 85, 88, 89, 92, 93, 94, 95, 96, 116, 117, 121, 122, 126, 127, 130
Perry sesquicentennial 103, 124
Perry Square 122
Philadelphia, PA 36, 52, 87, 96, 98
Pittsburgh, PA 11, 36, 52, 78, 97
Polk, President James 87
Porcupine 90
Port-au-Prince, Trinidad 85

Port Dover 53, 56, 75
Port Rowan 53, 55, 56, 74
Port Ryerse 53, 74, 82
Portsmouth, England 82, 83
Presque Isle 10, 13, 16, 35, 74, 103
Providence, RI 85
Put-in-Bay 49, 76, 85, 121

Q

Queen Charlotte 29, 90

R

Reed, Rufus 89, 90, 117
Reed, Seth 89
Reid, Robert 36
Revenge 23
Roosevelt, Franklin D. 103
Roosevelt, Theodore 96
Rybka, Captain Walter 14, 108, 112, 127
Ryerse, Amelia Harris 82

S

Salina 46, 47
Sandusky, OH 35, 62
Scorpion 61, 67, 69, 70, 72, 73, 83, 86, 90, 91
seasickness 21, 32, 36, 38
Seneca (Iroquois) Indians 25, 28
Sinclair, Arthur 45, 46, 52, 53, 55, 61, 62, 63, 66, 67, 68, 69, 76, 77, 86, 89, 91, 127
Sinclair, Upton 86
smallpox 25
Smith, Melbourne 109
Somers 90

ABOUT THE AUTHOR

David Frew is a visiting professor at Mercyhurst University in Erie, Pennsylvania, in the Graduate Organizational Leadership (MSOL) Program. He is a professor emeritus at Gannon University, where he served on the faculty and held a variety of administrative positions for thirty-three years; an emeritus director of the Erie County Historical Society; and president of his own management consulting business. His consulting clients have included Disney, GE Crotonville, several General Electric core businesses, Hammermill Paper Company, Lockheed Martin, Lord Manufacturing, NASA, FMC Technology, National Fuel Gas, the Regional Cancer Center, Hamot Health Center, St. Vincent Health Center, AMSCO/STERIS, the Second Harvest Food Bank and dozens of other regional, national and international clients.

Dr. Frew has authored or coauthored thirty-five books and more than one hundred journal articles, cases and papers. His work has appeared in publications ranging from refereed journals such as *The Journal of Applied Psychology* and the *Academy of Management Journal* to popular magazines including *Sail Magazine* and *Cruising World*. His nine books on Lake Erie ships, shipping and shipwrecks represent a narrative history of the industries that formed the backbone of the Great Lakes maritime economy.

He was born in Erie and graduated from Gannon. After a five-year career in engineering, he took a teaching fellowship at Kent State University, where he earned a doctorate in organizational behavior in 1970. His love of Lake Erie brought him back to Erie, where he became a faculty member and the founding director of Gannon's MBA Program. His career at Gannon included administrative posts as director of the Health Services Administration and Public Administration Programs and dean of the Graduate School. He retired from Gannon in 2003 and took over the executive directorship of Erie County Historical Society, where he was able to integrate his management and leadership skills with his love of Lake Erie history. In 2009, he returned to academia at Mercyhurst University.

Dr. Frew is married with three grown children and eight grandchildren. He is an avid racing and cruising sailor.

Visit us at
www.historypress.net